101 AMERICAN FOSSIL SITES
you've gotta see

101 AMERICAN FOSSIL SITES
you've gotta see

ALBERT B. DICKAS

2018
Mountain Press Publishing Company
Missoula, Montana

© 2018 by Albert Binkley Dickas
First Printing, January 2018
All rights reserved

Photos © 2018 by author unless otherwise credited.

FRONT COVER PHOTO
Microraptor, on display at the Dinosaur Center in Wyoming. Note the faintly visible feather imprints (grayish lines) on the front and rear limbs of this winged dinosaur.

BACK COVER PHOTOS, TOP TO BOTTOM
Shark teeth from Green Mill Run, North Carolina, courtesy Jayson Kowinsky
Sweet gum leaf from Clarkia Fossil Bowl, Idaho, courtesy Bill Rember
Trilobite from Penn Dixie Fossil Park, New York, courtesy Tasha Mumbrue
Petrified Forest, Arizona, courtesy National Park Service/Andrew Kearns

Library of Congress Cataloging-in-Publication Data

Names: Dickas, Albert Binkley, 1933- author.
Title: 101 American fossil sites you've gotta see / Albert Binkley Dickas.
Other titles: American fossil sites you've gotta see
Description: Missoula, Montana : Mountain Press Publishing Company, 2018. | Includes bibliographical references and index.
Identifiers: LCCN 2017051226 | ISBN 0978878426812 (pbk. : alk. paper)
Subjects: LCSH: Fossils—United States—States. | Paleontology—United States—Popular works.
Classification: LCC QE746 .D53 2018 | DDC 560.973—dc23
LC record available at https://lccn.loc.gov/2017051226

PRINTED IN CANADA

P.O. Box 2399 • Missoula, MT 59806 • 406-728-1900
800-234-5308 • info@mtnpress.com
www.mountain-press.com

I dedicate this volume to my grandchildren:
Emma Margaret
Gabrielle Nicole
James Douglas
William Elias
Kathryn Adrienne
Thomas Christian
Jessica Louise
Braden Patrick
May they grow to appreciate the wonders of nature.

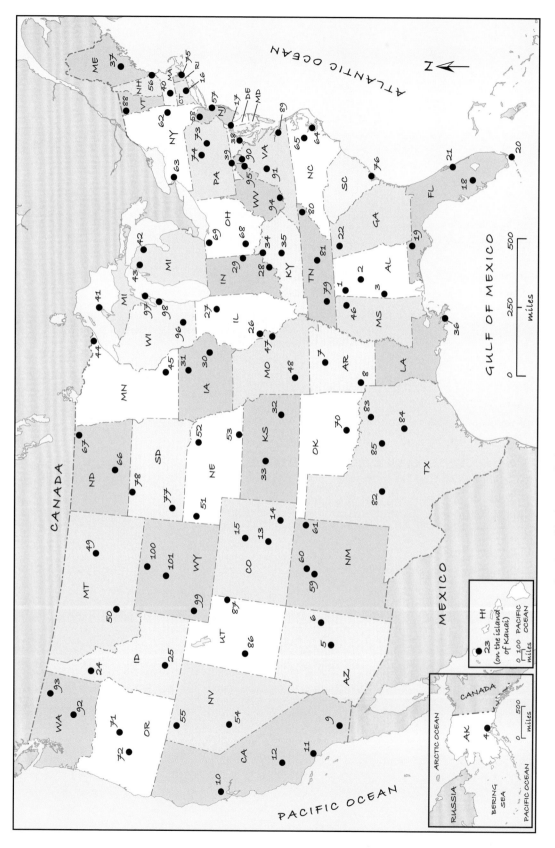

Locations of fossil sites in the book. The numbers correspond to specific sites.

Contents

PREFACE ix

ACKNOWLEDGMENTS xi

INTRODUCTION 1
   The Annals of Paleontology 1
   Fossils and Fossilization 4
      Body Fossils 5
      Trace Fossils 6
   A Short History of Life on Earth 6
      Hadean Eon: 4,600 to 4,000 Million Years Ago 7
      Archean Eon: 4,000 to 2,500 Million Years Ago 7
      Proterozoic Eon: 2,500 to 541 Million Years Ago 8
      Cambrian Period: 541 to 485 Million Years Ago 8
      Ordovician Period: 485 to 443 Million Years Ago 9
      Silurian Period: 443 to 419 Million Years Ago 9
      Devonian Period: 419 to 359 Million Years Ago 10
      Carboniferous Period: 359 to 299 Million Years Ago 10
      Permian Period: 299 to 252 Million Years Ago 10
      Triassic Period: 252 to 201 Million Years Ago 11
      Jurassic Period: 201 to 145 Million Years Ago 11
      Cretaceous Period: 145 to 66 Million Years Ago 12
      Paleogene Period: 66 to 23 Million Years Ago 12
      Neogene Period: 23 to 2.6 Million Years Ago 13
      Quaternary Period: 2.6 Million Years Ago to Today 13

FOSSIL SITES
1. Double Roadcut, Alabama 16
2. Minkin Footprints, Alabama 18
3. Moscow Landing, Alabama 20
4. Coyote Lake, Alaska 22
5. Indian Gardens, Arizona 24
6. Petrified Forest, Arizona 26
7. Devils Backbone, Arkansas 28
8. Marlbrook Marl, Arkansas 30
9. Anza-Borrego Desert State Park, California 32
10. Petrified Forest, California 34
11. La Brea Tar Pits, California 36
12. Sharktooth Hill, California 38
13. Marsh-Felch Quarry, Colorado 40
14. Picket Wire Track Site, Colorado 42
15. Denver Museum of Nature and Science, Colorado 44
16. Peabody Museum of Natural History, Connecticut 46
17. Chesapeake and Delaware Canal, Delaware 48
18. Bone Valley, Florida 50
19. Florida Caverns State Park, Florida 52
20. Windley Key Fossil Reef 54
    Geological State Park, Florida
21. Brevard Museum of History 56
    and Natural Science, Florida
22. Tibbs Bridge, Georgia 58
23. Makauwahi Cave, Hawaii 60
24. Clarkia Fossil Bowl, Idaho 62
25. Hagerman Fossil Beds National Monument, Idaho 64
26. Grafton Quarries, Illinois 66
27. Mazonia-Braidwood State Fish 68
    and Wildlife Area, Illinois
28. Falls of the Ohio, Indiana 70
29. Whitewater River Gorge, Indiana 72
30. Devonian Fossil Gorge, Iowa 74
31. Fossil and Prairie Park Preserve and Center, Iowa 76
32. Round Mound, Kansas 78
33. Sternberg Museum of Natural History, Kansas 80
34. Big Bone Lick State Historic Site, Kentucky 82
35. Danville Bryozoan Reef, Kentucky 84

36. Grand Isle, Louisiana   86
37. State House, Maine   88
38. Dinosaur Park, Maryland   90
39. Sandy Mile Road, Maryland   92
40. Beneski Museum of Natural History, Massachusetts   94
41. Horseshoe Harbor, Michigan   96
42. Lafarge Fossil Park, Michigan   98
43. Petoskey State Park, Michigan   100
44. Gunflint Chert, Minnesota   102
45. Wangs, Minnesota   104
46. W. M. Browning Cretaceous Fossil Park, Mississippi   106
47. Mastodon State Historic Site, Missouri   108
48. Riverbluff Cave, Missouri   110
49. Hell Creek Country, Montana   112
50. Museum of the Rockies, Montana   114
51. Agate Fossil Beds National Monument, Nebraska   116
52. Ashfall Fossil Beds State Historical Park, Nebraska   118
53. Little Blue River Outcrop, Nebraska   120
54. Berlin-Ichthyosaur State Park, Nevada   122
55. Virgin Valley, Nevada   124
56. Odiorne Point State Park, New Hampshire   126
57. Big Brook Preserve, New Jersey   128
58. Hamburg Stromatolites, New Jersey   130
59. Battleship Rock, New Mexico   132
60. Ghost Ranch, New Mexico   134
61. Clayton Lake State Park, New Mexico   136
62. John Boyd Thacher State Park, New York   138
63. Penn Dixie Fossil Park and Nature Reserve, New York   140
64. Aurora Fossil Museum, North Carolina   142
65. Green Mill Run, North Carolina   144
66. Heritage Center and State Museum, North Dakota   146
67. Pembina Gorge, North Dakota   148
68. Caesar Creek State Park, Ohio   150
69. Sylvania Fossil Park, Ohio   152
70. Arbuckle Anticline, Oklahoma   154
71. John Day Fossil Beds National Monument, Oregon   156
72. Lava Cast Forest, Oregon   158
73. Montour Preserve, Pennsylvania   160
74. Red Hill, Pennsylvania   162
75. Cory's Lane, Rhode Island   164
76. Edisto Beach, South Carolina   166
77. Mammoth Site, South Dakota   168
78. Petrified Wood Park, South Dakota   170
79. Coon Creek Science Center, Tennessee   172
80. Gray Fossil Site, Tennessee   174
81. Rock Island State Park, Tennessee   176
82. Fluvanna Roadcut, Texas   178
83. Ladonia Fossil Park, Texas   180
84. Waco Mammoth National Monument, Texas   182
85. Mineral Wells Fossil Park, Texas   184
86. U-Dig Fossils Quarry, Utah   186
87. Wall of Bones, Dinosaur National Monument, Utah   188
88. Chazy Fossil Reef National Natural Landmark, Vermont   190
89. Chippokes Plantation State Park, Virginia   192
90. Hogue Creek, Virginia   194
91. Mint Spring, Virginia   196
92. Ginkgo Petrified Forest State Park, Washington   198
93. Stonerose Interpretive Center and Eocene Fossil Site, Washington   200
94. Glen Lyn Roadcut, West Virginia   202
95. Lost River Quarry, West Virginia   204
96. Blackberry Hill, Wisconsin   206
97. Newport State Park, Wisconsin   208
98. Two Creeks Buried Forest, Wisconsin   210
99. Fossil Safari, Wyoming   212
100. Red Gulch Dinosaur Track Site, Wyoming   214
101. Dinosaur Center, Wyoming   216

GLOSSARY   218

REFERENCES   225

INDEX   241

Preface

Every now and then I interrupt my day-to-day meditations and think back to that long-ago decision to change my undergraduate major from history to geology. With academic advisor approval on file and a pristine copy of *Physical Geology* in hand—purchased at a cost of $4.75, a suggestion of just how long ago that decision was made—I attended the first class. Fifty minutes later, I left the lecture hall and stepped into a lifelong career.

Eight in the morning is an unsavory time for Monday, Wednesday, and Friday classes on any subject, but I never missed a lecture. Standing 6-foot-3 inches tall and clad in a natty sport coat complete with leather elbow pads, the instructor expounded not only on the wonders and processes of the physical world but also on his participation in two of Admiral Richard E. Byrd's expeditions to Antarctica in the years prior to World War II. Reminiscing on one particular episode, a 77-day, 862-mile trek by dogsled across icebound Marie Byrd Land—at the time the most ambitious research journey ever attempted in the polar regions—he recalled his orders: map the geology and conduct an intensive search for paleontologic specimens.

Never having heard of Marie Byrd Land or the term *paleontologic*, I raced to the university library for answers. Eureka! The lecture included a review of the search for fossils in a section of the southernmost continent so remote that even today it is considered the largest unclaimed territory on Earth. I was hooked—line and sinker. If one could gain a reputation of note, and make a living at the same time, by jaunting about the globe in pursuit of no-one-has-ever-been-here discoveries, I had to be the first among my peers to explore the possibilities.

Four years later I was enrolled in graduate school with teaching responsibilities for a course titled Historical Geology. Because the bedrock underlying the campus was nationally recognized as a Late Ordovician *lagerstätte*—a fossil site of extraordinary interest—I quickly discovered that an occasional field excursion to rock exposures overburdened with troves of brachiopods, horn corals, and trilobites was a student-applauded diversion from a steady program of classroom lectures. In short, I learned that fossils had attention-holding value.

Being assigned to the Gulf of Mexico division of a major oil company in the mid-1950s was an opportunity of historic significance. The "GOM" was still in its first decade of growth toward its status as one of the most important oil and natural gas provinces in the world. As one of the few offshore geologists associated with these pioneering years, my duties were twofold: a primary task of keeping up-to-date maps showing the ever-changing relationship between geologic structure and hydrocarbon production, and a secondary responsibility of investigating regional trends in hydrocarbon potential. The first assignment was, with experience, relatively easy because the physical nature of the Miocene-age formations varied little over distances of several miles, but the second, conducted over distances of more than 100 miles, was often impossible. Where analyses by physical character failed, however, paleontologic appraisal proved to be valuable. When, for example, *Bigenerina*, *Discorbis*, and *Marginulina* were employed as index foraminifera to trace Miocene Epoch strata across extended subsurface stretches of the Gulf, worthwhile results could be obtained. Very quickly, I gained working-world knowledge that fossils had economic value.

Until an unusually thick and unexpected stratum of shale was discovered in the Sacramento Valley of California in the early 1960s, the literature of fossil submarine channels was practically nonexistent. I lacked a prototype model to guide my studies, and the evaluation and geologic history of the 2,000-foot-deep, 50-mile-long Paleocene-age structure was the type of challenge I had not yet experienced. Fortunately, the rock filling the buried trench contained a suite of foraminifera—the same fossils used in Gulf of Mexico analyses—associated with a defined marine environment. Once this relationship was realized, the answers became evident, and I learned that fossils had value in determining species habitat.

One of my first impressions upon arriving in the Lake Superior area in the mid-1960s, to begin a three-decade-long academic career, was that I would have little involvement with fossils of any type. After all, the youngest basement rocks in the region were 1.1 billion years old, and they were bookended by even older material. My surprise was genuine, therefore, when I learned that just the year before a benchmark paper had been published documenting the discovery of microscopic, lifelike rods, spheres, and filaments in flinty outcrops located not more than 100 as-the-crow-flies miles from my campus office. Within a decade the 1.9-billion-year-old Gunflint Chert fauna was documented as the oldest preserved evidence of life in the world, and the generic names *Animikea, Gunflintia,* and *Eosphaera* were being uttered with reverence wherever two or more paleontologists gathered. I learned that fossils were not exclusive products of the post-Precambrian world—the dogma long held by scientists of diverse reputations; rather, they could very well exist wherever there are rocks to harbor them and people to recognize them. Today the remains of 3.5-billion-year-old cellular microbes encased in rocks in the outback districts of Africa and Australia are accepted as the oldest known fossils, and scientists are searching for even older life in every nook and cranny of the world.

For more than six decades I have pursued the wonders and mysteries of the physical world, as both an industrial geologist and a general-practitioner academic geologist. Throughout these years, as reviewed above, paleontology has been one of the essential tools in my how-can-I-better-understand-this-situation kit.

When offered the opportunity to author this travel guide of significant fossil sites throughout the United States, I readily accepted, knowing well the challenges that lay ahead. Now, four years, thousands of highway and airline miles, and far too many "5 snacks for $5" field lunches later, the journey is complete.

The simple fact that you are reading these words is an indication that you are also a fossil enthusiast. Why not then turn to the next page—and I encourage you to not stop there—and to follow along in my footsteps? Hopefully your travels will give you as much pleasure and insight into the intriguing "abyss of time" world of paleontology as I have experienced and learned to love.

Albert Binkley Dickas
Brush Mountain, Virginia
October 2017

Acknowledgments

Planning, field evaluating, and writing a travel guide highlighting locales where fossils can be observed or collected in all fifty of the United States cannot be accomplished without the cooperation and support of many individuals.

Steven J. Uchytil, senior geological advisor with the Hess Corporation, reviewed the text with the exactitude and precision of the corporate mind. His recommendations brought a needed sense of clarity to the final product. My editor at Mountain Press Publishing, James Lainsbury, has been generous in encouragement and excessive in patience. Having worked with him in the production of two prior geologic travel guides, I still marvel at his diplomatic ability in using his electronic "blue pencil" to create a higher degree of consistency in my use of dates and facts.

Many thanks go to my field-assistant daughter, Dr. Virginia D. (Ginny) Ford, for her persistence in probing wayside outcrops in Texas and Oklahoma in search of fossil treasure. Schooled in matters different from those of geology and paleontology, she raised many why, when, and what inquiries that enlightened my insight into how to better write this guide.

One individual has earned special recognition. My valued and esteemed friend Rachael M. Garrity was there almost every mile of the way, from Bangor to Key West and from Phoenix to Fairbanks, making sure every last item of site-specific information had been recorded, many "just one more" photographs had been taken, and GPS directions to the next site visit were on hand. She was there, along with my teenage grandson Thomas, when a South Dakota storm of golf ball–sized hail reduced our sport-utility vehicle from one-owner, well-maintained, 15,000-miles status to certified total wreck within the course of six minutes. Had the accompanying tornado-speed winds impacted the vehicle at a right angle, rather than head-on, it is conceivable this field guide would have forever remained a work in progress.

From conception to completion, I have been involved with this project for four years. Thousands of air and ground miles have been traveled, many potential sites visited, and 101 chosen for inclusion. I was familiar with the geology and paleontology of many; others I approached with a list of what, how, and, when questions; and a few I visited for the first time. At each locale I sought out the resident principals, eager to probe their expertise. In the ensuing months, I often followed up with them by telephone or email with a list of additional questions and needs. With rare exception, the enthusiastic responses I received were the light that shone during those times when the tunnel seemed darkest. Their count is in the hundreds. To each and every one—park rangers, museum curators, state and federal survey personnel, university staff, and a diverse group of professional and amateur fossil collectors—I say a heartfelt thank-you.

Several final statements need to be made. I am responsible for each and every photograph shown without attribution in this field guide. Any mistakes that occur in the text are mine, and mine alone. When visiting any particular fossil site, please be aware of nearby traffic, respect private property, and observe all signs and safety warnings. You can obtain additional information concerning many of the fossil locations by visiting the site's affiliated web page. Finally, I should note that the sites I chose for this gotta-see guide differ in their appeal and presentation. Some are collection locales, others photographic sites. Some are fee based, others free. A few are of historic interest only. All, however, are in my opinion worthy of a visit.

EON/ERA		PERIOD		EPOCH	AGE*	FOSSIL SITES	DEVELOPMENTS IN THE HISTORY OF LIFE
PHANEROZOIC	CENOZOIC	QUATERNARY		HOLOCENE	today	21, 23, 72	
				PLEISTOCENE	.0117	9, 11, 18, 20, 34, 47, 48, 56, 76, 77, 83, 84, 98	humans evolve
		NEOGENE		PLIOCENE	2.6	9, 10, 18, 25, 65, 80, 89	
				MIOCENE	5.3	9, 12, 18, 24, 51, 52, 55, 64, 65, 71, 80, 89, 92	sharks become dominant in the sea
		PALEOGENE		OLIGOCENE	23	71	first whales, evidence of land mammals returning to sea
				EOCENE	34	4, 71, 93, 99	
				PALEOCENE	56	3, 4, 19, 78	mammals diversify
	MESOZOIC	CRETACEOUS			66	3, 8, 17, 33, 38, 46, 49, 50, 53, 57, 61, 66, 67, 79, 82, 83, 101	← EXTINCTION dinosaurs proliferate; first flowering plants first true birds
		JURASSIC			145	13, 14, 16, 50, 87, 100, 101	first birdlike dinosaur
		TRIASSIC			201	6, 40, 54, 60	← EXTINCTION reptiles give rise to dinosaurs; first mammals
	PALEOZOIC	PERMIAN			252		← EXTINCTION reptiles diversify
		CARBON-IFEROUS	PENNSYLVANIAN		299	2, 5, 27, 32, 59, 75, 85	spore-producing land plants flourish first reptiles
			MISSISSIPPIAN		323	1, 7, 81, 94	numerous crinoids and brachiopods
		DEVONIAN			359	28, 30, 31, 39, 42, 43, 62, 63, 69, 73, 74, 90, 95	← EXTINCTION jawed fish diversify; first amphibians
		SILURIAN			419	26, 97	coral reefs; first vascular land plants
		ORDOVICIAN			443	29, 34, 35, 37, 45, 58, 68, 70, 88, 91	← EXTINCTION life diversifies; includes trilobites, coral, brachiopods, nautiloids, crinoids, graptolites, bryozoans, and jawless fish, the first vertebrate
		CAMBRIAN			485	22, 58, 86, 96	hard-bodied life evolves; trilobites dominate
PRECAMBRIAN	PROTEROZOIC				541	41, 44	traces of soft-bodied life
	ARCHEAN				2,500		stromatolites formed by cyanobacteria
	HADEAN				4,000		no evidence of life
					4,600		

*Dates from the 2015 ICS International Chronostratigraphic Chart

Introduction

Science is often bedeviled by questions impossible to answer with precision, but the age-old query "How many trees are there in the world?" is no longer one of them. The answer is a smidgen over 3 trillion, as determined by a two-year-long analysis of 421,529 forest-density plots scattered across fifty countries. Should 3,000,000,000,000 be an incomprehensible figure, then consider that approximately the same number of seconds exist in the span of 94,638 years. In short, Earth is shaded by a "whole lot" of trees.

Beyond the subject of tree counts, one might ask how many species of life-forms exist today? Biologists have catalogued around 1.5 million and report some 15,000 new types each year. Supposedly, once all the undiscovered species are rounded up, the count will be 8.7 million species, give or take 1.3 million. Other estimates, however, list numbers as low as 3 million and as high as 100 million.

Paleontologists have had long-held debates on how many species have lived through the eons of geologic history. Some estimates exceed 5 billion, with the caveat that at least 99.9 percent are extinct. Whatever the estimate, the simple fact remains that following death most forms of life either decay, are consumed, or are destroyed in one fashion or another, leaving no record whatsoever that they ever existed.

Perhaps the most pertinent inquiry of all—at least within the context of this field guide—deals not with the quantity of prior life but with the evidence of prior life. Locales exist where fossils are as common as dandelions in an ill-kept lawn, but they are the exception, not the rule. Barring those magical moments when all the elements of success converge, fossil discovery is largely a matter of luck and persistence, and fossil rarity is perhaps the very element that makes paleontology such an interesting avocation.

Once called toadstones, devil's toenails, and dragon's teeth, fossils are the biologic linkages that join deep time with tomorrow. Fossil presence alone, however, could not bring enlightenment to centuries of superstition and disbelief; that took the pioneering work of individuals who were able to recognize curiosities of nature as true evidence of preexisting life. The life stories of fossils are, in many ways, the story of life itself.

The Annals of Paleontology

When did humans first become curious about past life-forms? A 20,000-year-old collection of pierced Jurassic snails discovered in a Cro-Magnon burial site outside the village of Les Eyzies, France, in 1868, may be evidence of a dawning curiosity, and the skeletal remains found with them may be those of the first paleontologist. No one knows his name, or if he even had one. Many writers have noted other instances of early humans attaching special importance to fossils. For example, the writings of twelfth-century philosopher Moses Maimonides describe using fossil sea urchin spines as an antivenom for snake bite; and a 15,000-year-old pierced Silurian-age trilobite found at Arcy-sur-Cure, France, was supposedly worn as a prized amulet. By the time Charles Darwin's benchmark publication *On the Origin of Species* was released in 1859, *paleontology*—a word coined in 1822 by Henri Marie Ducrotay de Blainville, editor of the *Journal de Physique*—had become a widely recognized field of study, albeit one of contentious debate.

The word *fossil*, attributed to the Latin *fossilis*, meaning, in a broad sense, "anything dug up," was early on applied to a wide variety of stone and stonelike objects without any consideration of their origin. In the fifteenth century, Leonardo da Vinci was one of the first investigators to recognize that certain undefined objects were the preserved remains of ancestral life-forms, and that their presence in Italian hillside strata was proof that ancient seas had formerly inundated southern Europe.

Much early controversy in the field was fueled by the seventeenth-century Irish bishop James Ussher, when he proclaimed that Earth had been created on Sunday, October 23, 4004 BC, a declaration based on genealogical information contained in his Bible. Soon, divergent schools of thought were in full flurry: one believed that Earth was but 6,000 years old and fossils were the remains of organisms that had perished during Noah's flood, and others, dating back to the early decades of the eighteenth century, that Earth was much older and fossils were evidence that life had been evolving for a very long period of time.

In his 1665 publication *Micrographia*, English philosopher Robert Hooke compared a hand-sized specimen of oak wood with a specimen of "petrify'd wood" and concluded the latter had formed by "having lain in some place where it was well soak'd with petrifying water . . . water as is well impregnated with stony and earthy particles." His choice of words is one of the earliest published records describing a process of fossilization whereby living matter is transformed and thus preserved in the geologic record.

After dissecting the head of a huge shark caught off the coast of Italy in 1667, Nicolas Steno, a Danish-born contemporary of Hooke, was surprised to find that the teeth bore an uncanny resemblance to "tongue stones" found in relative abundance along local beaches. Roman author Pliny the Elder had once described these strange stones as objects having fallen from either the sky or the moon. Ignoring prevailing opinion, Steno declared them to be fossil shark teeth.

Steno's future work led to the shaping of one of geology's bedrock principles, the *principle of superposition*: in a vertical sequence of undisturbed rock the oldest layer is at the base and the youngest on top. This simple declaration gives a relative sense of age not only to sequences of layered rock but also to any fossils contained within the rock. Though scientists such as Hooke and Steno were developing the basic premises of paleontology, the belief that Noah's flood had affected much of what could be seen in the natural world still dominated the thinking of many of Europe's most prominent scholars.

During the *age of reason*, the eighteenth-century philosophical movement that championed the notion that enlightened judgment could free humanity from the misconceptions of unreasonable dogma and religious authoritarianism, fossil collecting became popular both as hobby and vocation. In France, Georges Cuvier became passionately interested in the trend, and in 1796 he published a paper comparing the anatomy of modern elephant bones to mammoth fossils. One concluding sentence stands out with its unexpected wording: "All of these facts, consistent among themselves, and not opposed by any report, seem to me to prove the existence of a world previous to ours, destroyed by some kind of catastrophe" (Rudwick 1997). With little fanfare Cuvier, today acknowledged as the father of paleontology, gave the scientific world two new and startling concepts to contemplate and debate: *extinction*, the disappearance of a species so that it no longer exists anywhere, and *catastrophism*, the doctrine that extreme fluctuations in Earth's flora and fauna are caused by disasters of exceptional impact, followed by the creation of different organisms. The scientific advancement of paleontology was on the right path, but forks in the road caused continued controversy.

Meanwhile, across the channel, an Englishman was putting his interests in fossils to practical use during day-to-day work as a surveyor and consulting engineer. William Smith learned the tricks of his trade the hard way, spending years in the field investigating the specific physical character and fossil content of the bedrock underlying the moors and bogs of the entire kingdom—"each stratum contained organized fossils peculiar to itself" (Phillips 2011). In essence, he was formulating the *principle of faunal succession*: particular fossil assemblages succeed one another in a regular and predictable order even in widely separated geologic formations. When his field notes became too bulky to conveniently handle, Smith was encouraged to consolidate them into an orderly scheme of mapping. In 1815 he finished the task, creating the first geologic map of Britain, colored by hand and reduced to a scale of 5 miles to 1 inch. This seminal accomplishment, and his later work *Strata Identified by Organized Fossils*, is why Smith is known today as the father of English geology.

At about the same time Smith was making a name for himself, Mary Anning was born into the family of an impoverished cabinetmaker in the south of Great Britain. As a child she supplemented the family income by selling marine fossils collected from Jurassic-age rocks exposed along the English Channel. She discovered a 17-foot-long skeleton of *Ichthyosaurus*, and several years later the remains of the first *Plesiosaurus*, events that gave widespread notice that a woman could be scientifically effective in a man's England. Initially, however, Georges Cuvier raised a ruckus by doubting the validity of the plesiosaur, but when the find was legitimized by the scientific community, Mary's skill was validated. The British Society for the History of Science deemed her the "greatest fossilist the world ever knew," and Prussian naturalist Ludwig Leichhardt named her the "princess of paleontology." Perhaps Anning's most unconventional contribution to science was the suggestion that bezoar stones, sometimes found in the abdominal cavity of ichthyosaur skeletons, were in reality fossilized feces. The English paleontologist William Buckland agreed with her and named the "stones" *coprolites*.

Meanwhile, philosophical chaos continued to disturb the scientific community as "old Earth" advocates increasingly challenged the "young Earth" school of thought that Bishop Ussher had founded sixteen decades earlier. Catastrophism still swayed the minds of some of Europe's most distinguished thinkers until 1830, when British-born lawyer and geologist Charles Lyell popularized the theory of uniformitarianism that James Hutton had presented nearly four decades earlier in his groundbreaking tome *Theory of the Earth*. This understanding, that the forces of nature in operation today differ neither in kind nor energy from those of the past—that is, "the present is a key to the past"—revolutionized both geology and paleontology, gave Earth an age rooted in deep time, and opened the minds of many to the validity of evolution. Science was awakening to the concept that Earth was indeed quite old, and as the nineteenth century got underway, even further strides in the advancement of paleontology were being made.

As a boy, John Phillips accompanied his engineer-uncle William Smith, famous for publishing the first geologic map of Britain in 1815, about the English countryside in pursuit of new geologic information. A shared interest in rocks and fossils was established, with the result that in 1841 Phillips published the first global geologic timescale that identified rock strata according to the types of fossils they contained. Three eras of relative time were introduced: Paleozoic, Mesozoic, and Cenozoic, each identified with distinctive breaks in the fossil record. Soon, named episodes of geologic time were specifically associated with the evolution of a particular class of fauna: for example, Cambrian time is known as the Age of Trilobites, and Devonian time as the Age of Fishes. The division between Earth studies and fossil studies, once distinct, was becoming more and more a gray area. Scientists in both Europe and North America were merging their mutual interests in deep time, rocks, and ancient life into one all-inclusive study—that of geology.

By 1822 the word *paleontology*, meaning the study of fossils and ancient life-forms, had been coined. At about the same time, fossil devotees on both sides of the Atlantic Ocean were becoming aware of the merits of developing natural history museums, repositories of fossils and fossil information that supported investigators as they attempted to unravel the mysteries related to the history of Earth and the evolution of life. This was also the heroic age of science, a time when most of the remaining geologic time episodes of the Phanerozoic Eon were being defined: eons were subdivided into eras that, in turn, were subdivided into periods, and then into epochs. Consider the classic case of this refinement process, which involved Roderick Murchison and Adam Sedgwick, both hopeful of being the first to define the early Paleozoic rocks of England. Comparing their notes, they found that Sedgwick's proposed late Cambrian Period, a division based on the physical characteristics of rocks, overlapped with Murchison's suggested early Silurian Period, which he based on fossil analyses. Unable to resolve the controversy, the onetime friends became enemies, and scientists debated the disputed boundary for nearly half a century until Charles Lapworth, a noted English authority on graptolites (extinct marine organisms), ordained the strata in question to be Ordovician in age, a solution considered so reasonable and convenient on the basis of general geologic logic that it was almost immediately accepted.

Following the American Civil War, the westward expansion of frontier communities, military outposts, and railroads further opened the door to fossil discovery. The heralded and unexpected unearthing of *Triceratops*, *Allosaurus*, and *Stegosaurus* fossils led to the birth of *dinomania*—the obsession with anything dinosaur. These were also the years of the infamous bone wars, a rivalry between two American paleontologists that paralleled the earlier disputes of Sedgwick and Murchison in England.

In many ways Othniel C. Marsh and Edward D. Cope had similar personalities: they were both determined, colorful, and vain in their quest for paleontological supremacy. Marsh was prone to theft, specimen destruction, and bribery, whereas Cope sought publication notoriety in a journal he purchased and controlled. Onetime friends, they became bitter antagonists when Marsh suggested that Cope had mistakenly reversed the bones of a plesiosaur skeleton by placing the skull at the tip of the tail. For twenty years they financed expeditions to the American West, each hoping for the fossil discovery that would lead to public and professional acclaim. By 1899 both of these titans of paleontology were dead, their finances and reputations all but bankrupt, but the fossil inventories they left became the basis for several outstanding museum collections. The bulk of Marsh's collection was willed to Yale University and the Smithsonian Institution, whereas most of Cope's thirteen thousand specimens were sold to the American Museum of Natural History.

Twentieth-century developments in science had a tremendous influence on the field of paleontology. Radioactivity, discovered in 1896, quickly became a tool scientists could use to assign fossils an absolute date and to determine rates of species evolution. In 1915 German geologist Alfred Wegener advanced the theory of continental drift, which posited that sections of the Earth's crust slowly drifted on top of a liquid interior. The theory was upgraded and reintroduced a half century later as plate tectonics, the concept that explains not only the distribution of earthquakes, volcanic activity, and mountain ranges, but also the global geographic spread of fossil flora and fauna. In 1980 the father-and-son team of Luis and Walter Alvarez identified an iridium-rich layer of 66-million-year-old clay and hypothesized it was evidence that dinosaurs had been obliterated by an asteroid impact. With that discovery and theory, dinomania became a worldwide subject of interest for a whole new generation of fossil enthusiasts.

Similar and dissimilar developments have elevated paleontology from an obscure field only discussed in traditional journals of academia to one of sexy headlines splashed across the colorful pages of newspapers, magazines, and websites everywhere. Analyses of the stomach contents of the mummified remains of Ötzi the Iceman, found in 1991 in the Italian Alps, gave the world insight into 5,200-year-old eating habits. *Jurassic Park*, the 1993 Hollywood production depicting a sabotaged theme park beset by raging, genetically engineered dinosaurs, quickly became the highest grossing film of its time. Four years later, the Field Museum of Chicago submitted the winning bid of $7.6 million for the fossil remains of Sue, the largest and most complete *T. rex* ever found. Today, an unprecedented wave of fossil-collecting mania is underway, financed by fossil-fuel-rich sheikhs and hedge-fund moguls enamored by the lure of possessing deep-time fossils and the accompanying bragging rights.

Paleontology has traveled a long, often contentious, sometimes confusing, but generally interesting road since that Cro-Magnon fossil collector was buried in France. Its future is rife with possibility: Will cloned trilobites once again rule the seas? Will *Archaeopteryx* fly through the air on raven-like wings? Will mastodons graze forests once again? Stay tuned, as the history of paleontology is far from over.

Fossils and Fossilization

A precise, acceptable-by-everyone-concerned definition of *fossil* does not exist. Derived from the Latin *fossilis*, meaning "something dug up," the word has been defined in many different ways by many different authors and many different paleontologists. Computer and literature searches yield a myriad of individualistic meanings, as seen in the wording of a half dozen examples:

- Fossils are any remains or evidence of life preserved in a geologic sense.
- Fossils are incomplete remnants of life that formerly lived.
- Fossils are remains or traces of some part of the anatomy of once-living things that are older than our recent geologic experience.
- Fossils are the remains or evidence of preexisting life.
- Fossils are the evidence of life that has been preserved by natural processes in the Earth's rock layers.
- Fossils are the evidence of life during past geologic time.

Perhaps the *Glossary of Geology*, fourth edition, says it most succinctly: "Fossil—loosely, any evidence of past life." The same source defines *paleontology* as "the study of life in past geologic time, based on fossil plants and animals and including phylogeny, their relationships to existing plants, animals, and environments, and the chronology of the Earth's history."

Fossils are uncommon critters, indeed. The chances that any individual specimen of life that ever existed, since the dawn of life more than 3.5 billion years ago, has been preserved is one in a million to one in a billion, depending on the circumstances. Organisms that die on land are rarely fossilized; what remains that scavenging animals and insects do not consume, bacteria and decay do. Burial in a mucky anoxic environment, such as a stagnant lake or the inhospitable depths of an ocean, however, enhances the possibility of preservation. Two major types of fossils exist: body fossils and trace fossils. Both represent the remains of living organisms.

Body Fossils

As the name suggests, a *body fossil* is distinctive evidence, in the form of bones, shells, and teeth, or molds and casts, that an organism existed. The latter two are unusual examples because they are representations of the organism rather than remnants. The dissolution of a rock-confined shell or bone creates a cavity—that is, a *mold* bounded by the external expression of the fossil. A *cast* is what it sounds like: a reproduction of the former fossil that forms when sediment or minerals fill the mold, or impression, created by dissolution of a rock-confined shell or bone.

In very rare instances, body fossils are found in toto, preserved with minimal or no change from their living state. Freezing, drying, and entrapment in tar or plant resin can preserve actual body tissue. Examples include extinct woolly mammoths of the ice age that occasionally weather from the frozen tundra of Siberia; tar-black mummified "bog people" cadavers of the Iron Age, complete with skin and hair, found buried in peat deposits throughout western Europe; and perfectly preserved, three-dimensional, 15-million-year-old flowering plants that once grew in a humid, tropical forest in the Dominican Republic. Complete examples of invertebrate body fossils are relatively common, including gastropod, brachiopod, cephalopod, and foraminifera shells; crinoid columns; bryozoan and coral remains; spicules from sponges; and the exoskeleton molts of trilobites.

Body fossils, however, are more commonly found in various degrees of alteration. The alteration of remains is generally divided into one of three categories—carbonization, permineralization, or petrifaction—depending on the extent to which the original material, size, shape, or structure of the organism has been modified.

Fossilized organic tissue and chitinous skeletal material, such as the soft parts of fish, reptiles, and marine invertebrates and the leafy parts of plants, are often preserved by *carbonization*, a process that produces films of residual carbon through decay and the vaporization of the volatile elements of buried organic matter. A carbon impression is left in the host rock outlining the structural design of the original specimen, often in considerable detail.

Material of a permeable nature can be preserved by *permineralization*, the deposition of inorganic substances, such as calcium carbonate, silica, and iron, into the pores of bones and shells by groundwater. This form of fossilization preserves the original shape of the tissue or organism, but the composition of the fossil is different, and it is heavier than the original and less susceptible to future destruction because of the deposition of dense material.

In a similar process, called *petrifaction* or *mineralization*, groundwater completely dissolves the original skeletal or bone material and deposits in its place another substance, such as calcite, pyrite, or silica. In this case, the resulting object is entirely altered to rock, and the resulting fossil is a *pseudomorph*, a "false form" of the original material. Wood is particularly susceptible to this form of fossilization.

The scientific value of any body fossil is heavily dependent on a few details: Exactly how old is it and precisely where and under what conditions was it found? While any fossil may be a specimen of curiosity to a collector, without definitive information it has no value for geologic mapping or the understanding of paleogeography, ancient climates, or the processes of evolution.

The first geologic map of Britain, published by William Smith in 1815, was mainly based on his ability to field identify a wide variety of invertebrate fossils, an expertise gained by years of experience surveying the countryside. Geologic mapmakers today use the same guiding logic, formalized by the *principle of fossil correlation*, the concept that any and all strata that contain assemblages of fossils that are all the same age must be of similar age to the fossils. When using this principle, geologists assume that extinct species never reappear in the geologic record and no two species are identical: the identification of either situation would make the concepts of faunal succession worthless in the same way two persons having the same fingerprint would make the Federal Bureau of Investigation's file of criminals relatively worthless.

Index fossils are particularly valuable to the application of the principle of faunal succession. These fossils, preferably unique and easily recognized, are geographically widespread and existed for a limited period of geologic time, meaning that their occurrence in a particular stratum allows geologist to finely delimit its age.

Throughout time organisms have adapted themselves to a variety of particular environments, including deltas, lakes, rivers, prairies, mountains, deserts, and the shallow and deep portions of the oceans. *Paleogeography*, the science of the distribution of ancient landmasses and seas in the geologic past, is heavily dependent on the presence and analysis of fossil remains; for example, corals, echinoderms, and shark teeth are evidence of former oceans, whereas petrified wood and woolly mammoth bones indicate the former presence of land. The absence of fossils could also indicate the prior existence of land.

Many forms of life, whether present today or found only as fossils, are dead giveaways for past climates. Fossil ferns collected on Alexander I Island on the Antarctic Peninsula prove the continent was much warmer in the past. Similarly, the presence of Cretaceous-age laurel and fig tree specimens in Greenland offer conclusive evidence that the world's largest island, now ice covered, basked under a subtropical climate some 80 million years ago.

Many species of microscopic, single-celled foraminifera swarm in the world's oceans, protected by exoskeletons they create by extracting calcium carbonate and two varieties of oxygen—the rare isotope oxygen-18 and the common isotope oxygen-16—from the seawater. In the 1950s scientists learned that the amount of each isotope a foraminifera uses to build its shell varies with the temperature of the water and thus with the prevailing climate: low levels of oxygen-18 indicate warm temperatures, and high levels correlate to cold conditions. By applying the oxygen-isotope process to the analyses of foraminifera shells contained in cores drilled hundreds of feet into seafloor sediment, they established a climate-revealing record of temperature change stretching back more than 300,000 years. Today the worldwide oxygen-isotope climate record extends back more than 100 million years.

Billions of humanoids have populated the Earth over the past 6 million years, but the earliest fossil remains have been reported for only about six thousand individuals. Some remains are complete skeletons, whereas other specimens comprise only a partial skull or a few teeth. Analyzed on an individual basis but considered together, they hold the clues that may lead to an understanding of how, and perhaps why, humans learned to walk upright and use tools, improved their quality of life, and adapted to different habitats and environments. In short, the study of human fossils is determining the story of the evolution of humankind.

The 3.5-billion-year-old fossil record documents many examples of the transition of one species to another over time. For example, studies of the extinct genus *Micraster*, a common sea urchin, the fossils of which are found by the hundreds of thousands in the Cretaceous-age chalk of Europe, show a trivial, albeit continuous, evolution over a 10-million-year period. As the organism acquired the habit of burrowing deeper into the soft sediment of its aquatic environment, its heart-shaped shell became more rounded, the creature's mouth moved steadily forward, and its breathing tube gradually extended outward so the creature could maintain contact with seawater.

Trace Fossils

Trace fossils, also known as *ichnofossils*, include a variety of rock impressions and other forms of evidence indicative of a range of activities, such as walking, running, resting, burrowing, eating, grazing, and defecating. The impressions can be in the form of scratches, bite marks, nests, tubes, trackways, footprints, grooves, skin imprints, root cavities, and material produced by organisms, such as coprolites—that is, fossil feces. Most trace fossils formed when organisms came into contact with soft sediment in the course of day-to-day activities. Fossilization takes place only when the impressions are covered with sediment before tides, winds, or storms destroy them.

Some trace fossils, such as the imprints formed by a dinosaur walking along a shoreline (also called a trackway), can provide impressive insight into the size and shape of extinct animals, including footprint shape and size, number of toes, whether the animal was two legged or four legged, and even body size and weight. Overall trackway patterns are useful in determining if the animal walked erect or moved in a sprawled, disorderly fashion suggestive of illness or being wounded. It is even possible to determine the relative velocity of the animal when it laid its tracks by mathematically comparing such measured factors as stride length, footprint length, and hip height.

Sometimes more of anecdotal curiosity than of scientific value, trace fossils can be almost three-dimensional and resurrecting in appearance. A noted European slab of sandstone contains the tracks of an insect being stalked by a lizard. The two sets of tracks slowly converge, and then the carnivorous predator walks alone. A flagstone from New England records the story of a drenched, tail-dragging dinosaur plodding through a wind-driven rainstorm of Triassic time; the even-paced stride suggests that the animal had little protection, such as vegetation, from the storm.

Trace fossils can also be very useful for identifying diet, behavior, and the environments in which animals lived. For many years paleontologists thought that the clusters of fish fossils in the Green River Formation of Wyoming were the result of either a calamitous shift to an oxygen-starved environment or that deepwater oxygen-poor environments prevented scavengers from consuming dead organisms. However, the discovery of feeding traces in the same rock horizon suggests that lack of oxygen was not the cause of death; rather, uniform rates of sediment deposition in a calm environment buried dead organisms before they were disturbed by the elements or scavengers. For a long time scientists believed that the deepest oceans were biologic vacuums because of their inhospitable pressures and temperatures. The discovery of trace fossils in rocks deposited in abyssal marine depths, however, provides evidence that the deep ancient seas may have teemed with life.

Curious, collectable, valued, and intensely studied, fossils are biologic Rosetta stones that record the many changes Earth and its inhabitants have undergone since life arose. Subject to numerous conditions that do not allow preservation and relatively rare in appearance, fossils offer clarifying insight into the misty chapters of deep time, and they serve as reminders that all things that are born must eventually die.

A Short History of Life on Earth

The historical account of life on Earth, dating back to around 4 billion years ago, is exceeded only by the history of Earth itself, beginning 4.6 billion years ago. The study of geology began some 250 years ago, and early on little consideration was given to the development of a chronologic timescale relating to such questions as "Did flora evolve before fauna" or "Is mountain range A older or younger than mountain range B?" As more and more rock units were identified and traced from one geographic area to another, however, the need to arrange information in a manageable form became apparent. The development of a relative timescale met this need.

Relative time is based on the comparative age relationship of differing rock units as determined by several basic geologic principles. One, already mentioned, states that in an undisturbed sequence of strata the uppermost unit is the youngest, another states that an intrusion of rock is younger than the rock it intrudes, and a third states that fragments of rock included in another rock must be older than the enclosing rock. The use of these principles by field geologists since the later decades of the eighteenth century led to the development of a timescale in which time is divided into relative divisions, beginning with the longest, the eon, which in turn is divided into eras, and then progressively into periods, epochs, and ages of time. Because these units are universally applied, geologists and paleontologists from different continents can effectively compare notes knowing, for example, that the Devonian Period is older than the Cretaceous Period. This relative scale is still being refined, with consideration being given to adding the Anthropocene Epoch, the time period during which human activity has been the dominating influence on Earth's environment.

The obvious next step in the development of a timescale was assigning absolute ages to not only rocks but also to the many subdivisions that composed the relative scale. The discovery of radioactivity in 1896 led to the development of an absolute timescale, whereby, for instance, it is not only understood that the Devonian Period is older than the Cretaceous Period, but it is known that the Devonian ended exactly 214 million years before the Cretaceous Period began. The absolute timescale is based on the fact that different radioactive elements have different rates of naturally occurring chemical transformation, and for any particular element the rate is constant and independent of changes in pressure and temperature, depth of burial, physical state, or association with any combination of other elements. This independence provides a reliable "atomic clock" by which the age of rocks and, by extension, the age of different episodes of geologic time can be quantified.

Historically, the science of paleontology was considered the study of life dating back to the Cambrian Period, beginning about 541 million years ago, when hard-bodied organisms first appeared in the geologic record. This interpretation changed, however, with the mid-twentieth-century discovery of more than twenty-five genera of soft-bodied fossils within Precambrian rocks in South Australia. (*Precambrian* is an informal term comprising the Hadean, Archean, and Proterozoic Eons.) This eye-opening and mind-boggling biota gave credence to the suspicion that life had been present during at least some of the more than 4 billion years that preceded the Cambrian. Today, these rocks that date back to an era when Earth's atmosphere had no free oxygen and its oceans were as steamy as a hot bath drawn for a field-fatigued paleontologist provide evidence of life.

HADEAN EON
4,600 to 4,000 Million Years Ago

Earth came into existence 4.6 billion years ago, born in a molten state from a rotating cloud of gas and dust—a hellish environment in which life as it is known today was impossible. Within the surprisingly short time period of several hundred million years, the surface of the planet changed from liquid to solid, and temperatures declined to the degree that the theoretical building blocks of nature were in place: an atmosphere, a hydrosphere, and a geosphere composed of a nourishing slurry of minerals. Life was, however, still not possible, as the atmosphere was a toxic mixture of methane, ammonia, carbon dioxide, and other gases, and the oceans contained a disastrous concentration of organism-destroying acids.

No evidence of Hadean life exists. In fact, to date no definitive Hadean-age rock has been found. Rocks collected along the shore of Hudson Bay in 2008 were found to be somewhere between 4,280 to 3,800 million years old, the former the oldest age for any rock reported from any region on Earth. Considering the age range, however, it is questionable as to whether these specimens should be considered late Hadean or early Archean.

ARCHEAN EON
4,000 to 2,500 Million Years Ago

Because fossils are destroyed by the extreme pressures and temperatures that form metamorphic and igneous rocks, geoscientists consistently look for them in sedimentary rocks. The oldest-known sedimentary rocks are associated with the Isua sequence of volcanic strata in southwestern Greenland. Because these 3.7-billion-year-old rocks were affected by extreme conditions of heat and pressure, no fossils have been found to date; however, they contain faint chemical signatures providing evidence that some form of life may have existed at this time.

The oldest documented assemblage of fossils is an eleven-species suite of organic-walled microbes found in the Apex Chert, a unit of the 3,465-million-year-old Pilbara Group of rocks in northwestern Australia. These simple and very tiny—about 0.0002 inch in diameter—microfossils are arranged in necklace-like strings of cyanobacterial filaments. They are proof positive that life rambled onto the evolutionary stage at about the midpoint of the Archean Eon. *Cyanobacteria* are single-celled, asexual, algae-like organisms that obtain their energy through photosynthesis, the process of converting light energy into chemical energy, with oxygen released as a waste product. Their rise to prominence is the reason Earth's original atmosphere slowly converted from a toxic to an oxygen-rich state, favoring the evolution of complex life.

Stromatolites are the most common category of Archean fossils, constructed by cyanobacteria that formed spongy, layered mats in shallow marine lagoons. Thin coatings of sediment collected on the glue-like surface of each mat, through which a new layer of algae grew. Repeated over and over, this process created cabbage-like to columnar structures, some many feet tall. They are the only megascopic fossils found in rocks that are 3.5 billion to 635 million years in age.

Earth entered the Archean Eon a seething and rambunctious planet and left it somewhat subdued. Life, the animated, indispensable state of vitality and being, coexisted now along with air, water, and rock. Earth would never again be the same.

PROTEROZOIC EON
2,500 to 541 Million Years Ago

Through the decades of the nineteenth and twentieth centuries leading up to the years of World War II, finding Precambrian fossils was the number one item on the to-do list of many paleontologists. Intensive searches were conducted wherever there was the possibility of finding such organisms, but the quest seemed almost hopeless. A new era of information and analyses, however, was underway with the serendipitous discovery, in 1946, of soft-body imprints in sandstone of the Ediacara Hills of South Australia. These fossils became the key to understanding that the evolutionary trends during Precambrian time had been far more sophisticated and extensive than formerly thought.

Fully 42 percent of the historical account of Earth is associated with the Proterozoic Eon, a time of mind-boggling change. Consider the topics of this eon: atmospheric conversion, global glaciation, mass extinction, and the invention of sex. A planetary midlife crisis was underway.

Rapidly rising levels of atmospheric oxygen during early Proterozoic time triggered a life-threatening signal to the simple, asexual organisms that populated Earth: change your habits or die. Many found this "oxygen catastrophe" overwhelming and thus expired, but others adapted by forming symbiotic relationships with each other. The new social order was composed of organisms that were on average ten times larger than their predecessors, sported a distinctive nucleus, and exchanged their genes sexually, so that each generation produced evolving genetic configurations. The sexual revolution was off and running.

Shifts in weather patterns created by the fragmentation of the supercontinent Rodinia in late Proterozoic time, along with reductions in greenhouse gases, caused global temperatures to decline to the point that glacial ice fields thousands of feet thick extended from pole to pole for millions of years. As much as 95 percent of aquatic life may have become extinct during this "Snowball Earth" phase of history.

By the end of the Proterozoic Eon, also the end of Precambrian time, Earth had experienced 88 percent of its history, the rock foundations of the protocontinents were complete, and North America had physically blossomed to nearly 75 percent of its present size. The road ahead was full of evolutionary promise—seemingly void of speed bumps and greased by the unparalleled effects of what is known as the Cambrian explosion.

—— PALEOZOIC ERA ——
CAMBRIAN PERIOD
541 to 485 Million Years Ago

Post-Precambrian time comprises three eras—the Paleozoic, Mesozoic, and Cenozoic, meaning "ancient life," "middle life," and "recent life," respectively—during which major changes in faunal composition took place. For the better part of 4 billion years the evolutionary processes of experimentation and change had produced a simplified lineage of minute, soft-tissue algae and bacteria—life-forms that left a nearly nonexistent fossil record. That was about to change.

The opening chapter of the Paleozoic Era was the Cambrian explosion, the "evolutionary big bang" that instigated a stunning variety of large, multicelled, skeletal life-forms. Representatives of practically every modern animal phylum living today made an appearance in a flurry of biologic innovation and diversification.

Not surprisingly, the early returns of this new-world initiative often resulted in bizarre life-forms that roamed the oceans equipped with masses of vertical spines, sets of five eyes, or mud-sucking, fire hose–shaped noses. All were intent on longevity. In time, however, the fogs of evolution evaporated and many wannabe life-forms became extinct. Alpha forms developed body plans designed to promote long-term survival; for example, marine fauna added protective, hard-shelled armor to existing cartilage and tissue.

Why the 30-million-year-long Cambrian explosion, beginning some 540 million years ago, occurred at all is a subject still under debate. Some paleontologists attribute it to sea-level rise and the subsequent increase in the global presence of continental shelves, oceanic habitats highly conducive to life, as well as increases in atmospheric oxygen levels, environmental changes that fueled energy-hungry innovations such as muscles, nervous systems, and the development of the tools of defense and combat. Others emphasize an influx of calcium and silica minerals into the oceans that supported the formation of exoskeletons, and the introduction of a new inventory of genes that controlled the processes of development and growth.

Whatever the causes, a cascade of events successively meshed, producing a vast array of marine life-forms. The new hard-body fossils became an integral part of the geologic record and a key to our understanding of the history of life on Earth.

ORDOVICIAN PERIOD
485 to 443 Million Years Ago

During the Ordovician, the second of seven time periods that make up the Paleozoic Era, a series of evolutionary experimentations resulted in an increased number of biologic families, from 150 to 400, and a threefold increase in the number of species. Filter feeders replaced seafloor mud grubbers, a transformation made possible by an increase in the amount of plankton, ubiquitous microscopic organisms that float in salt water. Hordes of trilobites cruised through colonies of coral, brachiopods, and crinoids that populated the warm, shallow seas. Graptolites proliferated, and bryozoans made their appearance and spread like modern-day crabgrass. Of all the geologic periods, many paleontologists consider the Ordovician as the one with the most dramatic increase in diversity.

This diversity is best seen in the fossil-rich strata exposed throughout the tristate area of Kentucky, Indiana, and Ohio. During the final decades of the nineteenth century, the long-term interest in these strata by an Ohio-based group of amateur fossil enthusiasts led to the founding of the Cincinnati School of Paleontology and to the development of the abundant and diverse catalogue of Cincinnati-area fossils that today is recognized worldwide as the official definition of Ordovician life.

Life during the Ordovician Period became even more varied with the introduction of vertebrates, menacing predators that cruised the oceans with fierce appetites. The earliest were jawless fish distinguished by a skeleton made of cartilage rather than bone. Finally, an additional evolutionary leap occurred with the transformation of Cambrian algae into early forms of plant spores that invaded moist terrestrial habitats.

The end of Ordovician time was marked by an evolutionary speed bump—a short-lived episode of glaciation that resulted in the extinction of more than 50 percent of all marine species.

SILURIAN PERIOD
443 to 419 Million Years Ago

Two principal events define the history of life during the Silurian Period: the appearance of coral reefs and land plants. Although the lineage of corals has been traced back to the Cambrian Period, they were extremely rare until Middle Ordovician time, almost 100 million years later. Early on, corals were scattered across the ocean floors in both solitary and colonial forms. During the Silurian, they began to cluster together to form reefs—massive wave-resistant, mound-like structures that grew above the surrounding contemporaneously deposited sediment. Sometimes called "rain forests of the sea" and most commonly associated with shallow tropical seas, reefs form the biologic backbone of some of the most diverse and productive environments on Earth.

As corals ever increased Earth's biomass, the second big evolutionary story was the development of the first unequivocal plants—branching, multicelled organisms that convert light energy into food and, in the process, produce oxygen. During the Silurian, plants developed surface cells that helped prevent desiccation and internal cells that provided stem support and a means of fluid transport, and the formerly bleak terrain was converted into a verdant landscape through the development of leaves. The "green revolution" that enveloped the land set the stage for an invasion of amphibians.

DEVONIAN PERIOD
419 to 359 Million Years Ago

By the dawn of the Devonian Period, jawless fish had been cruising the depths of the oceans for millions of years and were still abundant, but their anatomy and lifestyle were changing. Once sheathed in isolated bony scales and plates, they were now protected by body armor from nose to tail, an evolutionary advancement that increased their probability of reaching old age.

Placoderms—jawed fish that dated to the Silurian Period and had a primitive vertebrate skeleton—attained peak diversity, sharks made an appearance, and two groups of bony fish homesteaded the freshwater realm: the ray-fin, which became the dominant fish of the contemporary world, and the lobe-fin, which gave rise to land vertebrates that blazed the transition from water to land. The Age of Fishes was in full swim.

Onshore, primitive lowland forests inhabited wetland environments, evolving from 1-foot-high ground-cover shrubs to 90-foot-tall trees that housed a myriad of crawling and flying insects. Four-footed amphibians that had to adjust to problems of body support and dryland reproduction quickly exploited the new and canopied habitat.

Then, disaster in the form of extinction slowed the onward progress of life. Oceanic surface temperatures plummeted as much as 15 degrees, and some 75 percent of marine species were wiped out. Brachiopods and trilobites were especially hard hit, but the reef community was dealt the most severe blow. Cold-water-loving glass-sponge reefs largely replaced warmwater coral reefs, a substitution that suggests the onset of a global cooling event. A possible decline in the amount of greenhouse gas associated with the rise in terrestrial vegetation may have helped chill the atmosphere. Whatever the reason, it appears that glaciation was responsible for the extinction event. Life was expanding its hold on Earth by crawling out of the oceans and onto land, but death in the oceans defined the end of the Devonian Period.

CARBONIFEROUS PERIOD
359 to 299 Million Years Ago

The term *Carboniferous*, meaning "carbon-bearing," is used throughout Europe in reference to the large amounts of coal associated with deposits of this age. North American geologists, however, have long divided this episode of time into two subsystems based on rock content: older, calcareous formations of Mississippian age, and younger, coal-bearing strata of Pennsylvanian age.

In the United States the word *Mississippian* is practically synonymous with limestone, or, perhaps better said, crinoidal limestone. In reference to the dense, grassland-like thickets of long-stemmed "lilies of the sea" that flourished at this time, paleontologists labeled this period the Age of Crinoids. The Devonian extinction, which emptied the marine habitat for crinoid expansion, made this age possible.

Shaded by the canopies of crinoids, hordes of brachiopods covered the seafloor during the Mississippian, while corals declined in numbers. Graptolites became scarce, and trilobites were approaching extinction. However, the status of microscopic foraminifera, classed today as one of the most abundant organisms on Earth, changed from rare to common.

During the Pennsylvanian Period the focus moved from marine fauna to terrestrial flora. Forests of cycad, ginkgo, and coniferous trees, as well as meadowlands of horsetails, rushes, and ferns, flourished inland and in coastal swamps and marshlands. Sea-level fluctuations, near-riotous plant growth, and plant accumulations in oxygen-poor conditions left a legacy of more than one hundred seams of carbon-rich strata that fueled the industrial revolution in both Europe and North America. Waterlogged lowlands were also the ideal environment for both the shelled amniotic egg and the plant seed to develop, durable evolutionary advancements that allowed both animals and plants to reproduce on land, rather than offshore.

PERMIAN PERIOD
299 to 252 Million Years Ago

Headlined as the "Time of Great Dying," the Permian Period is intimately aligned with the greatest mass extinction event to have taken place in the 4.6-billion-year history of Earth. It was a time of unprecedented variance and unusual circumstances.

Shaped like the letter C, Pangaea stretched from pole to pole, a "one Earth" supercontinent topographically accentuated by three young mountain ranges: the Urals of Asia, the Alps of Europe, and the Appalachians of North America. Offshore, the continental shelf—the gargantuan fertility zone of the world's ocean—bordering the global Panthalassa Sea grew smaller, while onshore the climate changed from moist to dry with little seasonal fluctuation. The result was catastrophic: nearly every Paleozoic group of marine organisms was wiped out, and a full 70 percent of land species became extinct.

Gone forever were the trilobites and the blastoids and the rugose class of coral. Many species of brachiopods, ammonites, mollusks, and bryozoans were decimated, while crinoids faced a crisis of survival. Plants with offspring encased in seeds replaced fernlike shrubs, and the great coastal swamps of the Carboniferous Period dried up and were transformed into conifer forests.

Scientists have yet to find the definitive smoking gun to explain this unprecedented calamity, though they have investigated many causes: global warming resulting from a runaway greenhouse effect; global cooling resulting from glaciation; record-setting eruptions of lava in Siberia; the disruption of ocean currents following the amalgamation of Pangaea; and all of the above, working in concert to create a perfect storm of negative evolutionary and environmental conditions. Only one factor is free of suspicion: to date, no evidence of extraterrestrial impact has been found.

Whatever the cause, over a span of time as short as 60,000 years, a very significant number of the old cast of organisms were retired for good and a completely new lineup took their place. It was profound, it was cataclysmic, and it would take 10 million years to get the course of evolution back on track again.

MESOZOIC ERA
TRIASSIC PERIOD
252 to 201 Million Years Ago

When the curtain rose on the Mesozoic Era, the second of three geologic eras making up the Phanerozoic Eon, the stage was set for the arrival of a cast of four-legged land critters. They were obviously different from the yesteryear groupings of mobile and sedentary marine organisms. You can compare this transition to that of the big band era's melodic rhythms giving way to the raucous sounds of rock and roll—the Triassic was a time of turmoil and adjustment.

Pangaea was breaking up, proto–North America was approaching an equatorial position, and all existing polar ice caps had melted, with sea level rising accordingly. The area east of the Great Plains, however, was emergent and subject to the corrosive effects of erosion, a condition not favorable for the formation and preservation of fossils. In contrast, limestone was being deposited to the west, where a shallow sea provided an ideal environment for fossilization.

Those species that had survived the crisis of the Permian extinction quickly moved into the many vacated ecological niches. Different assemblages of snails, lobsters, crabs, and new species of fish and reptiles frequented aquatic habitats. On land, turtles, lizards, and crocodiles started their evolutionary journey toward the forms we recognize today.

Reptiles were now the dominant evolutionary force: one branch gave rise to land-dwelling dinosaurs, and another led to great marine populations of long-necked plesiosaurs, air-breathing mosasaurs, and dolphin-like ichthyosaurs. A few even took to the air. The pterosaur, with a wingspan approaching 36 feet, was the largest flying vertebrate to ever live.

Mammals also evolved at this time, although they maintained a subdued existence until the demise of the dinosaurs. The Triassic Period ended with yet another episode of extinction, but it paled in comparison to the great dying that had closed the Permian Period.

JURASSIC PERIOD
201 to 145 Million Years Ago

Enraptured by the storyline of the 1993 blockbuster movie *Jurassic Park*, many fossil enthusiasts have misperceptions regarding Jurassic life; for instance, the iconic trihorned *Triceratops* and powerful-limbed *Tyrannosaurus rex* did not evolve until the Cretaceous Period. Nevertheless, the Jurassic Period was a time of fearsome change. The Sundance Sea stretched across western North America from New Mexico to Alaska, temperatures and humidity were above normal, and vegetation growth rates approached those of kudzu along contemporary southern waysides. These factors led to an explosive increase in the number of organisms, especially that class of "terrible reptile" known to most children.

Two genera of dinosaurs dominated the Jurassic forests in an uneasy relationships of life and death: 25-foot-long, 2-ton, armored, and spiked vegetarians of the *Stegosaurus* genus, and 28-foot-long, 3-ton, flesh-loving members of the *Allosaurus* genus. Some paleontologists consider *Allosaurus* a top-of-the-food-chain predator that preyed on creatures its size and even larger, while others suggest these carnivores were opportunistic, feeding primarily on the sick, the young, and the elderly. Offshore, ichthyosaurs reached their population peak during Early Jurassic time, and some members of the plesiosaur group attained the status of the largest-known marine predator.

Beyond the land and the seas, the big news, however, was the arrival of *Archaeopteryx*, a magpie-sized flying wonder. Although its name is commonly affiliated with the phrase "first bird," paleontologists agree that this extinct vertebrate was actually a hybrid with both birdlike and dinosaur-like features. Like contemporary birds, *Archaeopteryx* sported wings with flight feathers, a birdlike beak, and a wishbone. In contrast, its dinosaur ancestry is evidenced by a handful of teeth and three claws that projected from the midpoint of each wing.

Jurassic life was, in short, impressive in variety, awesome in size, and ravenous in appetite, but it was only a sneak preview of the characters that abounded in the Cretaceous Period.

CRETACEOUS PERIOD
145 to 66 Million Years Ago

The phrase "Cretaceous extinction" is correctly associated with the word *dinosaur*, and for good reason. Miniature to humongous in size, earthbound to aquatic in habitat, vegetarian to predatory in appetite, these icons of the Mesozoic Era attained an astounding diversification of more than five hundred genera and then disappeared—seemingly overnight—66 million years ago.

During any single year of the Cretaceous Period, it is believed that no more than one hundred different species of dinosaurs wandered the countryside. Their anatomical differentiation over the course of their reign made for a zoo-like population of distinctive characters: the crow-sized, four-winged *Microraptor* was the smallest; the 120-foot-long, 100-ton *Argentinosaurus* was the largest-ever land animal; and the six-story-high *Sauroposeidon* wore the crown for being the tallest. Neither prowess, mass, nor menacing appearance, however, could protect them from the most infamous of the five great episodes of extinction that punctuate the pages of geologic history.

Extinction events are commonly associated with a litany of contentious causes; however, most paleontologists have long thought the Cretaceous extinction was caused by the effects of either meteorite impact or volcanic eruption. Both events would have filled the atmosphere with choking amounts of dust, blocked sunlight, inhibited photosynthesis, created lethal firestorms, produced sulfuric aerosols, and emitted invisible clouds of carbon dioxide. In short, both could have eradicated life.

Mandatory smoking guns—hard physical evidence—have been found for each event: the thick Deccan lava flows in India erupted during the last stages of the Cretaceous Period, and the Chicxulub crater, a 120-mile-diameter meteoritic footprint buried beneath the Yucatán Peninsula of Mexico, is thought to be 66 million years old. High iridium levels—specific chemical spikes associated worldwide with 66-million-year-old sedimentary rocks—are linked with each type of event as well, and iridium is common in meteorites and has been found in high concentrations in volcanic eruptions.

For a long time, investigators fell into one of the two hypothesis camps. Recently, an explanation conjoining both plausible stories has appeared in the pages of the scientific press, suggesting that the extraterrestrial impact in Mexico triggered a massive episode of volcanic eruptions in faraway India, creating a one-two knockout punch to the Age of Reptiles.

Whatever the culprit, the entire dinosaur line, with the exception of the branch that led to birds, and many other forms of life disappeared. As much as 85 percent of marine species and at least half of all land organisms simply ceased to exist. However, the mammalian class of vertebrates, the root of contemporary humanity, survived.

—— CENOZOIC ERA ——
PALEOGENE PERIOD
66 to 23 Million Years Ago

The instant of geologic history identifying the inauguration of the Paleogene Period, the first of three chapters in the Cenozoic Era, is one of the sharpest known in the annals of evolutionary history. For more than 150 million years, dinosaurs and mammals had lived in a predator-prey relationship, one dominating and the other cringing in the recesses of evolutionary change. With the demise of the dinosaurs, however, the curtain rose on the Age of Mammals.

New opportunities were made available to any organism capable of adapting a range of body styles and to newly developed environmental niches. Global sea-surface temperatures rose between 10 and 15 degrees Fahrenheit and then declined to normal levels. This climate fluctuation lasted for millennia, during which numerous populations of foraminifera were killed and the world of small vertebrates changed, from a population of rodent-like creatures to numerous genera containing nearly every modern order of mammal. The temperature fluctuation also transformed forests from dense woodlands filled with deciduous hardwoods to extensive grasslands of ferns and palms.

In many ways, the Paleogene fauna of central North America—horses, rabbits, dogs, peccaries, camels, rhinoceroses, and elephant-like mastodons—resembled that of contemporary East Africa's savanna. Fish filled the habitat voids created by the demise of the mosasaurs and plesiosaurs, but the big news was the arrival of whales, the evolutionary consequence of land mammals migrating offshore.

In western North America, the Laramide orogeny—the driving tectonic force behind the development of the Rocky Mountains—formed a series of massive north-south-oriented basins that today harbor a treasure trove of energy-producing resources. Troughs filled with carbon-rich shale in Colorado and Utah contain the equivalent of more than 2 trillion barrels of petroleum, many times the proven hydrocarbon reserves of the Arabian Peninsula, and one Wyoming trough alone contains an estimated 800 billion tons of Paleogene-age coal. Oil is created when fossil marine biomass is transformed by excessive pressure and temperature, and coal forms as fossil vegetative biomass is altered. Commonly and often confusingly referenced as "fossil fuels," these economy-driving resources result from the long-term chemical and biologic transformation of enormous concentrations of fossils that were deposited in marine and terrestrial environments.

By the end of the Paleogene Period, life-forms still with us today filled the oceans, dominated the land, and had taken to the air. The surface of Earth, both geographically and biologically, began to assume modern-day appearances.

NEOGENE PERIOD
23 to 2.6 Million Years Ago

Constituting a mere 20.4 million years—less than 0.5 percent of the 4,600-million-year history of Earth—Neogene time was one of continuity. Worldwide, the climate continued to cool and become drier, the great canopied forests that shaded the landscape continued to give way to grasslands, and the topographic relief of the land assumed new dimensions. Continents were in the final stages of geographic rearrangement, and major mountain ranges were born: the Pyrenees and Himalayan chain in Eurasia, and the Andes, Sierra Nevada, and Cascades chain in North and South America. Ice fields formed on the ever-rising mountain peaks; sea levels dropped in response; emerging land bridges connected the gaps that separated Africa from Eurasia and Eurasia from North America; and, over the course of ten million years, North and South America became linked by a chain of volcanoes.

These geographic changes resulted in a global epidemic of wanderlust as animal populations everywhere explored the expanding vistas of grasslands. New trends were thus established, characterized by migration rather than evolution, and the episode of colonization known to paleontologists as the Great American Biotic Interchange was inaugurated. Ground sloths, armadillos, porcupines, and opossums moved north from South America, while dogs, deer, bears, cats, camels, skunks, horses, llamas, and raccoons moved south out of North America. About the same time, elephant-like mammals and rhinoceroses crossed the land bridge from Eurasia into North America.

Offshore, undulating fields of kelp—a form of brown algae—created friendly environments for sea otters and dugongs, an herbivorous mammal. Sharks once again assumed a dominant role in the marine food chain, led by the biggest of them all, the 50-foot-long, 50-ton brute *Megalodon*. "Meg" teeth can reach lengths of more than 7 inches and have long been considered a "wow" fossil by shark enthusiasts.

QUATERNARY PERIOD
2.6 Million Years Ago to Today

The Quaternary Period, known both as the great ice age and the Age of Man, is the most recent—and shortest—episode of Earth history. It is subdivided into two epochs: Pleistocene and Holocene. The Pleistocene, 2.6 million years to 11,700 years ago, is associated with some thirty cycles of advancing and retreating glacial ice. In the northern hemisphere, ice sheets up to 2 miles thick extended from their birthing grounds around Hudson Bay south to the fortieth parallel. The footprints of these rock-gouging invasions are numberless, including U-shaped valleys, the five Great Lakes, the Finger Lakes of New York, moraines, and a countryside clothed with fertile soil.

As the continental-scale tongues of ice oscillated across the landscape, they impacted the habitat of resident fauna. Eventually, a long list of mammals, including saber-toothed cats, giant ground sloths, mastodons, mammoths, short-faced bears, and giant beavers, faced worldwide extinction. In North America, horses, camels, stag moose, and dire wolves ceased to exist.

Just as every coin has opposing sides, so too does the history of Pleistocene life. One side is the story of ice and extinction, and the other involves hominids emerging from the rift valleys of Africa millions of years ago. Standing erect and armed with an evolving kit of tools, they forged a way forward, exploiting every nook and cranny that provided food and safety. Today the population of *Homo sapiens* exceeds seven billion individuals—not bad for such a short evolutionary run.

The advent of Pleistocene ice field retreat, beginning some 12,000 years ago, initiated a period of global warming that peaked during the Holocene Maximum, 7,000 to 4,000 years ago, during which many of Earth's great civilizations flourished. The most recent global climate change is known as the Little Ice Age, a 300-year-long period of cooling that began around AD 1550 and may have been the genesis for the bubonic plague that ravaged Europe. Since then the Earth has become increasingly warmer. Along with associated sea-level and weather alterations, these climate fluctuations have affected the population distributions of both humans and animals. The distribution of the affiliated fossil record was similarly affected.

Many paleontologists today suggest that a third epoch of history should be added to the Quaternary Period. Called the Anthropocene, it is defined as the time during which human activity has been the dominant influence on Earth's climate and environments. If the name is sanctioned by the scientific community, the Anthropocene Epoch could well document a new era of exploration for fossil evidence on Earth and even beyond, into the deep, dark recesses of space. Regardless of "our" time's name, many more chapters on the history of life remain to be written.

FOSSIL SITES

1. Double Roadcut, Alabama

34° 35′ 19″ North, 87° 56′ 27″ West

Sea bud and corkscrew-like Late Mississippian–age fossils abound at this rural site.

Periods of continental-scale tectonic volatility followed by life-sustaining episodes of quiescence. A foreboding statement indeed, terse and economical in length, and when applied to terrain that would one day become the Heart of Dixie, it becomes a preamble to significant geologic change.

About 330 million years ago Alabama was located near the southern border of Laurasia, the massive landmass that was an amalgamation of the ancestral territories of Greenland, Europe, Asia, and North America. To the south lay Gondwanaland, a similarly sized continent that was erratically crushing its way north, slowly closing the Rheic Ocean, the body of water that separated the two landmasses.

In Late Mississippian time, Alabama's portion of Laurasia was inundated by a warm, shallow embayment, an extension of the tropical Rheic Ocean. As the landmasses conducted their slow-motion dance of collision—full-body continental impact followed by tectonic relaxation and geographic separation—the depth and shoreline position of the embayment fluctuated. During episodes of impact, erosion intensified in the developing highlands, embayment waters became turbid with sediment as sandstone and shale formations were deposited, and marine life was stressed by these adverse environmental conditions. In contrast, during periods of tectonic relaxation, the embayment waters were relatively sediment-free and charged with nutrients and oxygen, a wide variety of organisms flourished, and the shells of calcareous organisms were the dominant material deposited on the seafloor, eventually forming limestone. Today, both amateur and professional paleontologists recognize the Late Mississippian–age limestone formations of Alabama as outstanding sources of museum-grade fossils.

Exposures of Bangor Limestone are common throughout northern Alabama. The double roadcut, a readily accessible exposure along both sides of Alabama 247, 1.4 miles northeast of the Colbert-Franklin County line and 20 miles southwest of Tuscumbia, is chock-full of fossils, including brachiopods, horn corals, crinoids, gastropods, bryozoans, and the occasional shark tooth or trilobite fragment. The real and unexpected treasures, however, are *Pentremites* and *Archimedes*, two rather uncommon fossils.

Pentremites is a genus of blastoid, an extinct category of stemmed echinoderm, marine invertebrates related to modern starfishes, sea urchins, and sea cucumbers. Often called "sea buds" and "rosebuds," and looking very much like miniature hickory nuts, blastoids are treasured for their almost perfect and unusual radial symmetry. The typical animal lived within an ovoid shell constructed of fourteen interlocking plates of calcium carbonate, five of which functioned as food grooves, channels that directed food to the organism's mouth. Like their cousins the crinoids, they lived a sedentary life attached to the seafloor by a relatively short, root-bearing stem, feeding mainly on planktonic organisms. *Pentremites* is widely recognized as an index fossil of the Mississippian Period.

Archimedes is one of the most peculiar fossils known to paleontologists. Its corkscrew shape could easily be mistaken for an unwanted object swept from a mechanic's workbench. A type of extinct bryozoan, the genus is named after the Greek mathematician Archimedes, the inventor of the water screw. In reconstructed form, each *Archimedes* consisted of a central coiled stalk that supported a lacelike veil punctured by thousands of microscopic pores. Each pore was home to an individual bryozoan animal.

The double roadcut is the source of many Mississippian-age fossils. Those that appear to be miniature hickory nuts and minute screws, however, are worthy of special attention, for they are the ones to be labeled as "awesome," handled with reverence, and collected with lasting interest.

Mississippian-age fossils are easily collected from both the road-level rubble pile and the massive overlying Bangor Limestone (center) exposed along both sides of Alabama 247.

Along with the quirky, corkscrew-shaped Archimedes, *fossil enthusiasts can also find the delicate, webbed, and fan-shaped* Fenestella *bryozoan (below dime), often called "lace coral," encased in the Bangor Limestone.*
—Courtesy of Jim Lacefield

Hickory nut–like in appearance, these extinct Pentremites *lived anchored to the seafloor by a stemlike column of circular plates. Penny for scale.*
—Courtesy of Jim Lacefield

A cluster of horn corals in the Bangor Limestone. Quarter for scale.
—Courtesy of Jim Lacefield

2. Minkin Footprints, Alabama

33° 48′ 29″ North, 87° 09′ 59″ West

Well-preserved, abundant, and diverse trackway fossils of the earliest reptiles.

The task was considered impossible, but that didn't deter the first person from hiking the entire 2,168 miles of the Appalachian Trail, from Mt. Katahdin, Maine, to Springer Mountain, Georgia, in 1948. Nowadays more than two thousand wannabe thru-hikers attempt the task each year, but at most only 25 percent complete the journey. The successful few certainly deserve plaudits, but they should know they conquered the short version of the mountain chain.

About 300 million years ago, the long version extended from Alabama, through New England, across Newfoundland, old England, and on to the Highlands of Scotland, and ultimately the northern reaches of Norway. The Atlantic Ocean did not yet exist, and the Appalachian Mountains were not yet tectonically mature. Thru-hiking the Appalachian Trail in a single year back then surely would have been impossible—it was double in length and crossed scum-covered bogs, too numerous to count, that were lush with vegetation that eventually became Pennsylvanian-age coalfields. Hiking the Appalachian Trail today is as popular an adventure as ever, but alternate sources of energy are gradually becoming more significant than the Appalachian coalfields. However, one defunct strip mine was granted new life when a deep-time treasure, acclaimed to be the most important site of its age in the world, was discovered.

By the late 1990s, coal production from the Mary Lee seam of the Pottsville Formation had become an item of Walker County history. The Union Chapel Mine, located 30 miles northwest of Birmingham, was being environmentally reclaimed when the discovery of a trackway made headlines. The tracks were identified as belonging to a primitive four-legged reptile-like creature, a type of life never before found in Pennsylvanian-age rocks in Alabama. In 2005 the 35-acre locale was officially dedicated as the Steven C. Minkin Paleozoic Footprint Site, in memory of the individual largely responsible for its preservation.

Lithology tells the whole story of what the landscape looked like at the time: coal overlain by strata containing marine fossils and interbedded with shale housing both land-dwelling and aquatic organisms translates to a shoreline setting in the freshwater sector of a major tide-dominated delta. Body fossils are rare at the Minkin site, but trace fossils—tracks, trails, and other evidence of once-living organisms—abound. At least fifteen species of vertebrate and invertebrate species have been found among the thousands of catalogued specimens. There are trackways of crawling horseshoe crabs and prowling millipedes, insect feeding traces and larvae burrowings, small amphibian footprints, vestiges of fish schooling in shallow pools, and thousands of tetrapod prints. Paleontologists have interpreted the latter tracks as having belonged to a primitive 6-foot-long salamander-like reptile, reputed to be the largest terrestrial animal alive at the time.

The trace fossils occur with fossils of a variety of plants that grew in the swamps adjacent to the tidal flats and deltas: seed ferns, giant horsetails, and treelike cycads typical of a multistory forest ecosystem. Because they do not represent actual remains of preexisting animals, trace fossils cannot be used to create museum exhibits of the animals' actual appearance, but they are useful for determining how the animals moved, their size, whether they traveled in herds or individually, and what environments they preferred.

You can gain access to the Minkin footprint site by participating in periodic collecting expeditions sponsored by the Alabama Paleontological Society. The collecting area features a cliff—the working face of the former open-pit mine—surrounded by slabs of shale that house the fossils. Unless their finds are of museum quality, participants can keep what treasures they find, take-home representatives of the world that existed just before the dinosaurs emerged.

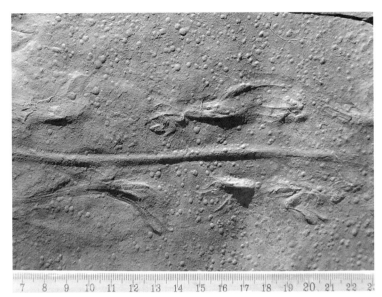

Tracks of the small reptile Cincosaurus cobbi, *the most common vertebrate trace fossil in the Pennsylvanian-age strata of Alabama. The prominent tail-drag impression (center) and raindrop impressions (small bumps) create a sense of a scurrying creature seeking shelter from an unexpected rainstorm. Centimeter ruler for scale.* —Collected by and courtesy of Ronald J. Buta

Impressions of U-shaped burrows in positive relief on the bottom surface of a shale bed. The faint, undulating ridges were probably made by sea foam. Centimeter ruler for scale. —Collected by and courtesy of Ronald J. Buta

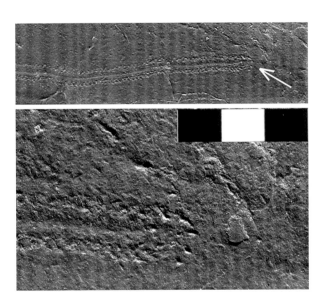

This millipede trace is unusual because the trackway abruptly ends (arrow in top view, amplified in bottom view). The fate of the scampering critter is a mystery; perhaps it was snatched from the mud by a vertebrate or a large flying insect. Centimeter ruler for scale. —From *Footprints in Stone: Fossil Traces of Coal-Age Tetrapods* by Ronald J. Buta and David C. Kopaska-Merkel, © 2016 by the University of Alabama Press, used with permission

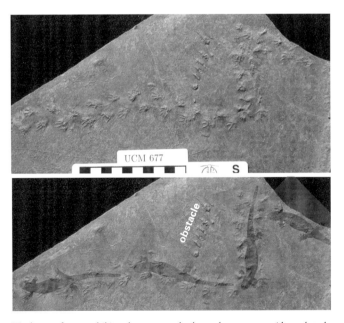

Trackway of an amphibian that apparently changed course to avoid an obstacle. The schematic illustration in the lower frame shows the animal walking around the obstacle and then continuing its right-to-left course. Centimeter ruler for scale. —From *Footprints in Stone: Fossil Traces of Coal-Age Tetrapods* by Ronald J. Buta and David C. Kopaska-Merkel, © 2016 by the University of Alabama Press, used with permission

3. Moscow Landing, Alabama

32° 25′ 55″ North, 88° 02′ 09″ West

Foraminifera fossils and episodes of humor and calamity make this site iconic.

The unpretentious brush-covered western bank of the Tombigbee River in the vicinity of Moscow Landing is home to a pair of episodes that share a common heritage: a history written in stone. The first incident, both trivial and amusing, occurred a mere century ago, whereas the second, involving a simultaneously catastrophic and evolutionary event, dates back 660,000 centuries.

By 1919 the Dixie Overland Highway, an early American automobile trail designed to link Georgia and California, had been completed, with the exception of a bridge spanning the Tombigbee River near Demopolis, Alabama. An unusual fund-raising scheme was devised whereby six hundred "roosters," identified as personages of the era, pledged money to construct the Rooster Bridge, a structure that carried traffic until 1980. At Moscow Landing—the riverfront termination of Sumter County Highway 25—a bullet-pocked gray granite monument bearing the names of eighty-eight donors, and the adjacent rusting remnants of the Rooster Bridge, are reminders of this curious episode from when the horseless carriage was transforming America.

The second incident is recorded in the chalky and sandy limestone and interbedded marl forming the shoreline of the Tombigbee River at Moscow Landing. The deposits stretch 1 mile downstream from the landing. This chapter of geologic history details an incident that brought an end to the Cretaceous Period, brought death to the dominant form of life on Earth, and opened the door to the Age of Mammals.

Although many suggest the Deccan Traps of India are a reason for the Cretaceous-age extinction, the worldwide scientific community is in general agreement that the Age of Dinosaurs ended abruptly when a 6-mile-diameter meteorite crashed into the Yucatán Peninsula of Mexico 66 million years ago, releasing a billion times the energy as that produced by the atomic bombs that fell on Hiroshima and Nagasaki in 1945. Forming a crater that measures 110 miles in diameter and 3,000 feet in depth, the impact produced a cloud of dust and soot that blocked sunlight for years and poisoned Earth's atmosphere with acid rain. The end result was the extinction of as much as 85 percent of all animal species on Earth, including the dinosaurs.

Physical evidence of this cataclysmic event can be found at Moscow Landing: minute, barely visible to the naked eye, glassy spheres called microtektites, and chunks of chalk torn from the Cretaceous shoreline by colossal tsunami waves created by the meteorite impact. The riverfront outcrop is composed of Late Cretaceous sedimentary strata of the Prairie Bluff Chalk overlain by Paleogene-age rocks of the Clayton Formation. Two diagnostic index genera—*Racemiguembelina* in the older strata and *Morozovella* in the younger material—have established the time relationship of these rocks. These fossils belong to the foraminifera class of unicellular, principally marine microorganisms that develop pinhead-sized calcareous shells. Industrial and academic scientists worldwide value them for their usefulness in analyzing the age and depositional environments of sedimentary rocks.

Standing at this paleontologically significant locale—known internationally as the K-Pg extinction boundary (*K* for Cretaceous and *Pg* for Paleogene)—during periods of low water, the visitor can firmly place one foot on rock identifying the Age of Dinosaurs, and the other on strata of the Age of Mammals. Moscow Landing may lack eye-catching megafossils, but the presence of tsunami debris and microtektites, and the sense that this is a deep-time site where life changed forever, distinguishes it from other well-researched Mesozoic-era fossil sites of the eastern United States.

Sometimes misidentified as fossils, the microtektites in this highly magnified view are actually small (0.04 inch or smaller) spheres of silicate glass produced when a massive meteorite struck sedimentary rock. —Courtesy of Jim Lacefield

From Rooster Bridge, look downstream to where light-colored strata of the Prairie Bluff Chalk (left) comes in contact with gray strata of the Clayton Formation (right), identifying the K-Pg boundary.
—Courtesy of Jim Lacefield

It is believed that tsunami waves tore the white chunks of chalk (foreground) from the Cretaceous seafloor and then redeposited them; they mark the uppermost units of the Prairie Bluff Chalk at Moscow Landing. The K-Pg contact lies at the base of the gray strata (upper left), and the remnants of Rooster Bridge are visible on the horizon.
—Courtesy of Jim Lacefield

4. Coyote Lake, Alaska

61° 44′ 30″ North, 148° 53′ 59″ West

Unusual terrestrial fossil hunting in the shadow of a serene, terrane-confined lake.

Alaska is a land often described in terms of superlatives too numerous to list, but a handful convey a sense of its terrestrial magnitude. More than twice the size of Texas, the so-named Last Frontier straddles five time zones; is home to both 20,310-foot-tall Denali, the highest peak in North America, and an oceanic cavity in the Aleutian Trench that bottoms out at 26,604 feet below sea level; is the source of the 1964 magnitude-9.2 earthquake, the largest ever recorded on the North American continent; and, in spite of global warming, is still the province of several Rhode Island–sized glaciers.

Generations of geologists have entered the Alaskan outback, hammer, hand lens, and aerial photographs in hand, intent on filling in one more sector of unmapped land with a description of the contained rocks and a résumé of geologic history. Even with these efforts, thousands of feet of rock spread over thousands of square miles of territory remain relatively unknown, owing to remoteness and the brevity of the summer field season.

Geologists have often conducted field mapping in Alaska by identifying terranes, rather than using the conventional means of identifying individual formations. A *terrane* is a body of rock bounded by faults and characterized by a geologic history that differs from adjacent terranes. This out-of-the-ordinary broad-brushed style of mapping does not address the many remaining mysteries of Alaskan paleontology, such as the relative absence of dinosaur fossil sites, even though the state is home to more strata of Mesozoic age (the Age of Reptiles) than any other state. One particular fossil site of opportunity, however, has been identified.

North of Anchorage, the Glenn Highway—also known as Alaska 1—advances eastward along the north bank of the Matanuska River, nestled between the Chugach and Peninsular terranes to the south and the Wrangellia terrane to the north. Paleogene-age plant fossils are found in the sedimentary rocks exposed along the 46-mile stretch of paved road connecting Palmer with the toe of the Matanuska Glacier. Weekend paleontologists often focus their explorations along tributary cut banks, but for the adventurer who desires a prime dig site, Coyote Lake is a location of opportunity.

Situated at the termination of Jonesville Road, 2.5 miles north of the blink-and-miss-it community of Sutton, football field–sized Coyote Lake is bedded in sandstone, shale, thin units of freshwater limestone, and bituminous coal–bearing strata of the Chickaloon Formation. Deposited at a time when the polar regions of Earth were much warmer, and in an environment saturated with peat-generating swamp forests, Chickaloon strata contain a multitude of wood, plant, and freshwater invertebrate fossils of Paleocene and Eocene age. Horsetail, sequoia, alder, elm, oak, magnolia, fern, pinecones, seedpods, and freshwater mollusks round out the inventory of catalogued specimens.

Though fossils can be found along the well-tramped south shore of Coyote Lake, the real prizes lie in the stair-stepped outcrops exposed in the scree slopes along the north side of the lake. Remnants of a 54-million-year-old forest are also found in the vicinity, including tree trunks more than 3 feet in diameter containing in excess of 150 visible tree rings. One massive upright specimen can still be seen along the west side of the lake.

Fossil opportunities abound at Coyote Lake, but a word to the wise: the last 0.2 mile of the Jonesville Road is a real test for vehicles without four-wheel drive, and it is essentially impassable when wet.

The scree slopes and outcrops along the north side of Coyote Lake harbor a surprising supply of leaf, fern, and petrified wood fossils of vegetation of Paleocene and Eocene age.

The carbonized remains of Chickaloon Formation plant fossils indicate that fields of vegetation thrived in a warm-temperate subtropical climate. Quarter for scale.

Several branching, secondary veins and the central midrib (between fingers on right) of a 54-million-year-old leaf are visible in this split sample of Chickaloon Formation coal.

5. Indian Gardens, Arizona

34° 19' 20" North, 111° 06' 40" West

Huge numbers of *Composita* brachiopods have long been collected at this well-known fossil site.

Paleontologists of all categories, from novice to professional, can become frustrated with the range of questions that come up when investigating terrain that has yet to feel the footprint of humanity. The potential for a reputation-making discovery increases with every step, but so does the distinct possibility of disappointment: Are the rocks too old to contain fossils? Has metamorphism destroyed any that had once existed? Has erosion swept the region clean of all evidence of preexisting life? If, however, one is not in search of that one elusive specimen for a collection and merely wishes to fill a ziplock bag with fossils in a short period of time, sites do exist where the pickings are as easy as plucking ripe fruit off a low-hanging branch. Formerly known as the Naco Paleo Site, Indian Gardens, about 13 miles east of Payson on the south side of Arizona 260, is one such locality.

The parking lot is big enough to accommodate a convoy of vehicles, and the zigzag entry is designed to slow down any children impetuously bent on declaring their first discovery. The site contains units of the Naco Formation, a sequence of ledge-forming limestone and gray slope-creating shale, deposited in Pennsylvanian time when Arizona was inundated by an inland sea. The formation hosts a variety of invertebrate fossils, including bryozoans, corals, crinoids, and gastropods, but the site is famous for *Composita subtilita*, a bivalve fossil found throughout the United States, as well as in Bolivia and Peru. This tear-shaped brachiopod is scattered like raisins in pudding in the ledge rock, and it is commonly found lying loose in the weathered slope debris.

Brachiopods are bivalves that first appeared in the geologic record in Cambrian time and still live today. Since contemporary specimens live exclusively in marine environments, their presence is convincing evidence that the host rock was deposited in a marine environment. A complete collection of fossil "brachs" would consist of more than twelve thousand species subdivided into some five thousand genera. Scientists use them as indicators of climate fluctuations that occurred during the Paleozoic Era. For instance, a small number of brachiopod species indicates small differences in temperatures between equatorial and polar regions; larger numbers of species developed during cold periods. They are also useful for evaluating contemporary environmental conditions. The decline of brachiopods living on the California continental shelf that began in the 1820s is attributed to the rise in water turbidity connected to the increasing number of cattle grazing onshore. More recently, one extant species is being employed to evaluate water quality changes that may result from offshore drilling operations in the Sea of Japan.

An identifying characteristic of an extant brach is the presence of a pedicle, a thin, fleshy appendage the organism uses to anchor itself to the seafloor. In fossilized brachiopods, fossil hunters can easily identify the pedicle valve by its pedicle opening, an aperture through which the living appendage protruded. The remaining shell is known as the brachial valve. The shells of youthful *Composita* are smooth, but with maturity growth lines accentuate them, as does a weak to strong undulating fold in the pedicle valve that corresponds to a form-fitting furrow in the opposite brachial valve.

Indian Gardens is the ideal location to spend an hour or so leisurely gathering the fossil remains of a fingernail-sized invertebrate that has survived the trials and tribulations of evolution for the last half billion years.

Specimens of Composita subtilita can be extracted from ledge rock of the Naco Formation (upper right), but whole unweathered examples can easily be found in the slope-forming debris beneath it.

Brachial valves of Composita subtilita. *Larger and more mature specimens display a defined and well-developed undulating furrow, or trough, a feature known as the sulcus. Quarter for scale.*

Pedicle valves of Composita subtilita, *each with a prominent pedicle opening. Penny for scale.*

25

6. Petrified Forest, Arizona

34° 48′ 55″ North, 109° 51′ 56″ West

Petrified trees preserved in strata deposited in a river system that was possibly larger than any known today.

Although commonly referenced as the Grand Canyon State, in the minds of many visitors Arizona is a land of sand, sun, and minimal rainfall. Meteorological facts support these impressions: sections of the state receive less than 3 inches of precipitation per year, and long-standing records show that Arizona is the sunniest of all US states, basking under the light of the sun more than 80 percent of the time. Scroll back through the chapters of deep time, however, and a different climate scene emerges.

About 225 million years ago Arizona was 10 degrees north of the equator, along the southwest edge of the supercontinent Pangaea. The landscape resembled that of present-day Costa Rica—humid, lush, verdant, and blanketed by forests. The curtain had recently dropped on the Permian Period, bringing an end to some 96 percent of all marine species and 70 percent of all terrestrial vertebrates—the greatest extinction event in Earth's history. The dawn of the dinosaurs was underway, and early forms of these "terrible" lizards populated the countryside in combative harmony with emerging species of crocodile-like reptiles.

Covered with ferns, horsetails, and cycads, the forest floor was shaded by galleries of ginkgoes and conifers, some of which grew to 10 feet in diameter and towered 180 to 200 feet in the sky. The land was drained by an Amazon-like, northwest-flowing river system that undercut the forest giants during floods and carried them downstream, denuding the trunks of branches and roots. While the majority of the trees eventually decomposed and disappeared, some clumped together as massive logjams, were buried in volumes of river sediment and blankets of ash spewed from distant western volcanoes, and became part of the geologic record.

Once the logjams were buried, time and water chemistry worked their magic, dissolving silica from the ash and depositing it in the logs' fibrous hollows and cell interiors, a process that preserved many exact details of the original surfaces and interiors of the trees. One petrified giant bears a finger-width burrow that interrupts its surface, the mark of an insect that ate its way into the tree.

Traces of iron and magnesium combined with quartz to embellish the now-petrified logs with a rainbow of colors—red, yellow, orange, purple, black, gray, and white. The final result of this multimillion-year episode of destruction and preservation is Petrified Forest National Park, a 229-square-mile, semiarid, grass-covered plain and badland region of northeast Arizona. It is acclaimed as one of the best locales in the world to examine the fossil record of the Late Triassic Period.

Twelve genera of petrified trees have been identified, the most prominent being *Araucarioxylon*, *Woodworthia*, and *Schilderia*, all recognized as extinct conifer-like trees that reproduced with seed-bearing cones. Clusters of broken logs are distributed across four well-marked trail regions within the park: Jasper Forest, Crystal Forest, Long Logs, and Giant Logs. The trees have weathered out of conglomerate of rounded pebbles, brown-and-white cross-bedded sandstone, and red-toned shale layers of the Chinle Formation. Radioactive age dating suggests the trees died between 218 and 211 million years ago.

Any visit to Petrified Forest National Park, which is open on a daily basis, should begin at the Rainbow Forest Museum, where exhibits of fossil wood and prehistoric vertebrates make this extinct sylvan world come alive.

Multihued, turned-to-stone remnants of a once-thriving forest litter the floor of Petrified Forest National Park, framed by badland topography sculpted from the Chinle Formation by wind and rain. Each log segment appears to have been cut by a saw, an effect created by silica, which breaks along clean lines. —Courtesy NPS/Andrew Kearns

Old Faithful, a 35-foot-long 44-ton root system, accentuates the midsection of the 0.4-mile-long Giant Logs Trail, located behind the Rainbow Forest Museum.

Because they are associated with a subtropical to tropical environment, the trees of Petrified Forest National Park probably grew year-round and thus did not form annual growth rings. The rings seen in this specimen likely formed during periodic droughts. Pen for scale.

7. Devils Backbone, Arkansas

35° 53′ 41″ North, 92° 36′ 54″ West

Dramatic fossil-bearing rocks of the Ozark Plateaus Province at a scenic roadside crest.

Novice paleontologists hoping for significant fossil discoveries in Arkansas should first become familiar with the state's *physiographic provinces*, regions characterized by similar geologic histories. An understanding of the interrelationship of Earth-altering events, rock type, and depositional environments that defines each region is essential to the successful discovery of prehistoric life treasures. Five such provinces make up the geography of the Natural State.

Sparse fossils are found in the rocks of the Ouachita Mountains Province because the deep ocean-basin setting in which they were deposited was nearly inhospitable to marine animals. Discovery success in the young and unconsolidated sediments of the Mississippi Alluvial Plain Province is also limited due to the ease with which erosion destroys fossils. Most rocks in the West Gulf Coastal Plain Province don't bear fossils, with the exception of limited exposures of Cretaceous age; in these strata collectors have reported finding dinosaur, shark, and marine reptile remains. Almost the entirety of the rock column of the Arkansas Valley Province is composed of Pennsylvanian-age sandstone, shale, and coal beds; these deposits are associated with brackish water, coastal depositional conditions that are generally favorable for the preservation of only plant fossils.

However, faunal Arkansas fossils are common in hillside bluffs, roadcuts, and abandoned quarries throughout the Ozark Plateaus Province. The sedimentary rocks of this province were deposited in continually changing oceanic environments during the Paleozoic Era, resulting in an invertebrate fossil–rich sequence of shallow-water reef and marine limestone rocks intermixed with freshwater coastline deposits.

One of the more accessible fossil-collecting sites in the Ozark Plateaus Province lies at the northern limits of the Boston Mountains, adjacent to the US 65 scenic overlook known as the Devils Backbone, located 1.5 miles south of downtown Marshall. Two formations of the Late Mississippian Period are exposed in this 40-foot-high buttress of rock: the Fayetteville Formation and the overlying Pitkin Limestone. At the time these strata were being deposited across northern Arkansas, the assembly of the supercontinent Pangaea was nearing completion. The amalgamation of the landmasses of Laurasia and Gondwanaland elevated the land and triggered a retreat of the long-standing sea to the south. This fender-bending tectonic collision resulted in a change of rock type: relatively deepwater, dark-gray limestone and black shale of the Fayetteville Formation gave way to shallow-water, lighter-colored beds of the younger Pitkin Limestone. The contact of these formations is quite noticeable along the east side of the highway, opposite the area with four canopied picnic tables.

Both formations contain a variety of 320-million-year-old invertebrate fossils. Crinoid segments, brachiopod valves, and specimens of *Archimedes*, a corkscrew-shaped bryozoan named after an early type of water pump supposedly invented by Greek mathematician Archimedes, are common. Extinct since the Triassic Period, *Archimedes* had a spiral curtain-like mesh wrapped around a corkscrew stalk that was anchored to the seafloor. Individual stalk examples and mesh examples are common, but specimens with the mesh attached to the central corkscrew structure are rare. Specimens of coral, gastropods, ammonoids, and blastoids are less prevalent here.

Fossils weather out relatively easily from the shale beds of the Fayetteville Formation and can thus be found in the rubble at the base of the Devils Backbone outcrop. In contrast, Pitkin fossils are tightly bound within this formation's limestone matrix, and the lack of color contrast between specimen and rock makes identifying fossils—at first glance—somewhat difficult.

Shaped like miniature life preservers, these hollow calcareous fossils are the disarticulated remains of the flexible stems of crinoids, or sea lilies. Dime for scale.

Rarely exceeding 1 to 3 inches in length because of their small diameter and delicate nature, Archimedes fossils can be found twisting in both a clockwise and a counterclockwise direction. Dime for scale.

Brachiopod fossils at Devils Backbone can be difficult to see because they are small and their color is close to that of the enclosing limestone. Dime for scale.

Time at Devils Backbone is best spent probing the ditch-filling rubble with a trowel searching for fossils that have weathered out of the massive beds of Mississippian-age limestone and shale.

8. Marlbrook Marl, Arkansas

33° 46′ 22″ North, 93° 54′ 18″ West

Fossil oyster shells are evidence that sea-level fluctuations have been, and always will be, the norm.

Science pundits have been spreading the word for decades, and while the majority of Americans are believers, a sizable group of doubters continue to ignore the indisputable facts. Tide-gauge records dating to the early twentieth century show that, whatever the cause or causes may be, ocean shorelines around the world are slowly being inundated. Some of the biggest invasions have been recorded along the east coast of the United States: since 1970 mean sea level has risen 5 inches at Boston and close to 9 inches at Norfolk, Virginia. This phenomenon has implications for both public safety—saltwater contamination of potable water aquifers and an increase in flood risks—and public perception. Should the ocean continue to rise along the coast of Miami at the rate it has since 1913, Florida's sun-and-fun reputation as the American Riviera may have to be changed to the American Venice, a less pleasing reference to the flooding that periodically devastates this fabled Italian city of canals.

The widespread presence of marine invertebrate fossils in outcrops of sedimentary rocks as old as 541 million years is proof positive that sea-level change has been a continuous North American event for a long time. Examples of sea-level change abound, but one in particular is illustrative of the entire geologic catalogue of continental oceanic invasions: the flooding of the ancestral Mississippi River Valley.

Following hundreds of millions of years of push-and-shove tectonic forces, the amalgamation of the supercontinent Pangaea was complete in late Paleozoic time. At this point, opposing stretch-and-pull forces began the process of landmass fragmentation, a change in global geography still underway today. As North America passed over one of the deep-seated thermal plumes housed in Earth's mantle, an elongated arch of land formed. Erosion combined with subsequent cooling and subsidence of the crust eventually produced a south-to-north trough that flooded during the Cretaceous Period as far north as southern Illinois. This tongue-shaped invasion of the ancestral Gulf of Mexico was the Mississippi Embayment.

Fossils eroding from exposures of Marlbrook Marl on both sides of Arkansas 355, 1.5 miles north of the community of Saratoga, are evidence for this ancient sea-level rise. Blue gray when freshly exposed and moist, and white to light brown when weathered and dry, the Marlbrook Marl is a chalky, highly plastic mixture of clay and calcium carbonate that was deposited across the southwestern portion of the Razorback State approximately 80 million years ago. Four notable species of oyster weather out of this 50-to-200-foot-thick, boot-sucking formation: *Gryphaea vesicularis*, *Exogyra cancellata*, *Exogyra ponderosa*, and *Ostrea falcata*. These extinct, benthic mollusks lived in shallow-water zones of the Mississippi Embayment, protected by calcareous bivalve shells that range in size from 1 to 6 inches. Fossil specimens are usually found in fragmented form, but intact examples can be found in low-lying ditches and creek bottoms. Reptilian mosasaur fossils are also found in this marl, reportedly once so abundant that locals used larger vertebrae as doorstops.

After reaching a high-water mark in the southern counties of Illinois 66 million years ago, the Mississippi Embayment began a slow retreat to the south. Eventually it was no longer an identifiable body of water. The Marlbrook catalogue of fossil oysters, however, remains as evidence that humankind will continue to be plagued by sea-level change, to the same disturbing degree as death and taxes.

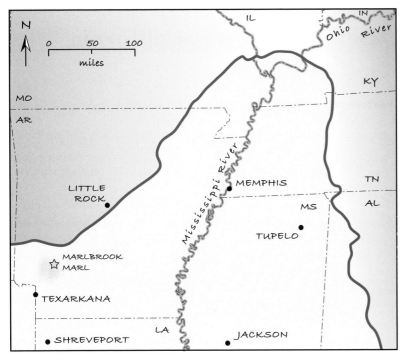

The maximum extent of the Mississippi Embayment in North America at the end of Cretaceous time, some 66 million years ago (blue). The Marlbrook Marl exposure in southwestern Arkansas is marked by a star.

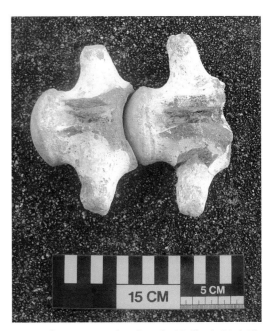

A pair of mosasaur vertebrae from the Marlbrook Marl. The ball-and-socket arrangement is suggestive of the flexible prowess employed by this marauder as it cruised the Cretaceous sea in search of food. Centimeter ruler for scale. —Courtesy of the Arkansas Geological Survey

Distinct ribbing representative of growth lines characterizes Exogyra ponderosa, *an extinct oyster found in a zone of Cretaceous-age outcrops that stretches from New Jersey to Mexico. Oyster is 6 inches wide.* —Courtesy of the Arkansas Geological Survey

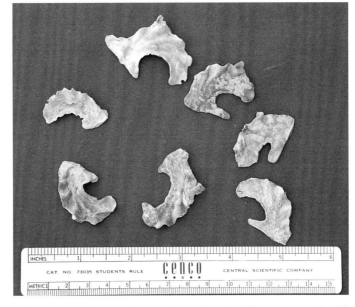

Differentiation of the more than one thousand species of Ostrea *is somewhat difficult, even for experts, but* Ostrea falcata *is one of the easiest species to identify due to its general U shape and the presence of seven to ten deep sets of ridges and depressions. Six-inch ruler for scale.* —Courtesy of the Arkansas Geological Survey

9. Anza-Borrego Desert State Park, California

33° 15′ 29″ North, 116° 24′ 24″ West

Once a savanna, this desert venue hosts a cornucopia of fossils.

Visitors to Anza-Borrego Desert State Park, about a two-hour drive northeast of San Diego, might be surprised to learn that evidence of the migratory meanderings of proboscideans—the order of vertebrates that includes mammoths and mastodons—have been found within the park boundaries at more than eighty localities. Strange but true, as the park lies within shouting distance of the western limits of the Sonoran Desert, a region noted for its thorny cholla cacti, creosote bushes, and lofty sand dunes—hardly an environment conducive to the well-being of elephant-like creatures. An enigma for sure, and one best understood by referring to a chapter of late Cenozoic Era history.

Some 20 million years ago, the expanse of the present-day southern California badlands was primarily an extended grassy savanna crisscrossed by rambling streams and shaded by an occasional riparian thicket of cottonwood and willow. The topography was heavily influenced by the Salton Trough, a steep-sided landward extension of the ancestral Gulf of California. Cloudburst-swollen rivers flowing into the trough—principally the Colorado, which was in the earliest stages of eroding the Grand Canyon—deposited loads of sediment. With each passing year, an expanding delta ever so slowly changed the pervasive marine environment into dry land. Analyses of fossil wood indicate the area was then annually drenched by four times the rainfall that dampens the region today.

To the south, tectonic maneuvers on a continental scale, and a long history of volcanic eruptions coincided to form the Isthmus of Panama 3 million years ago, a land bridge that allowed an almost unprecedented transfer of terrestrial and freshwater fauna back and forth between North and South America. These long-range migrations, known to science as the Great American Biotic Interchange, led to more than 550 types of plants and animals occupying and ultimately being preserved at Anza-Borrego. Many paleontologists consider this collection an unparalleled resource of international significance.

The catalogue of Anza-Borrego fossils includes specimens that range in size from microscopic plant pollen and algal spores to difficult-to-carry fragments of ground sloth, giant camel, bone-crushing dog, saber-toothed cat, a 6-foot-tall flightless bird, and a bathtub-sized tortoise. The remains of mammoths are the big attraction, cited as the most readily identifiable and abundant proboscidean fossils in the park. A 1.1-million-year-old skeleton of a mature female mammoth has "first lady" status. It is a nearly complete *Mammuthus meridionalis* specimen, considered one of the largest of all proboscideans and the earliest species of mammoth to exist in North America.

Spread across more than 600,000 acres of badlands and sandstone arroyos and accessed by 500 miles of road, Anza-Borrego is California's largest state park. It contains the longest continuous fossil record in North America, but collecting is prohibited; however, fossil-related interests abound. Beyond the "first lady" mammoth, a well-traveled 5-mile-roundtrip trail leads into the Carrizo Badlands, where a 12-to-15-foot-thick, 20-million-year-old oyster shell reef is exposed within the shadow of Elephant Knees Mesa. In places the dark-brown crust is so dense with shells and shell fragments that they appear to outweigh the entombing sandy matrix.

Academically inclined enthusiasts would do well to investigate the certification program conducted by the Anza-Borrego Desert Paleontology Society. Certified graduates can join caravan excursions into remote sections of the park, where they collect fossils and later clean, identify, and catalogue them under the guidance of experienced personnel. Whether conducted by vehicle, on foot, or as part of an organized group, fossil explorations within the Anza-Borrego badlands are experiences wholly unavailable anywhere else in the United States.

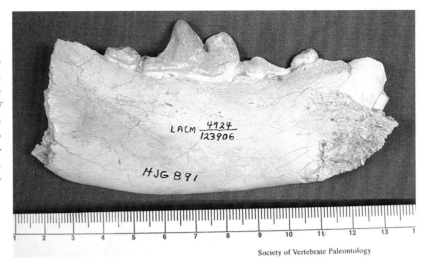

Jaw and teeth of Borophagus, *an extinct bone-crushing dog. Thought to have been a scavenger, this hyena-like creature fed on the marrow of bones left by other carnivores that occupied the Borrego Badlands. Centimeter ruler for scale.* —Courtesy of California State Parks, Stout Research Center, © 2016

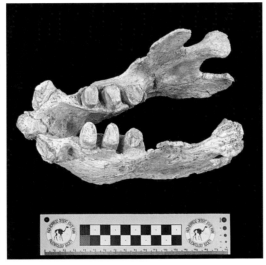

Anza-Borrego fossils of Megalonyx, *a 10-foot-tall ground sloth that ranged over North America as far north as Alaska. After examining bones found in a western Virginia cave, Thomas Jefferson proposed this genus name in 1797. Centimeter ruler for scale.* —Courtesy of California State Parks, Stout Research Center, © 2016

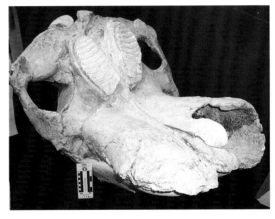

Upside-down view of the "first lady" mammoth skull, prominently displaying the molars. Ruler for scale. —Courtesy of California State Parks, Stout Research Center, © 2016

The dark-brown crust capping Elephant Knees Mesa is composed entirely of fossil oysters and other mollusks. Eroded shells cover the slope in the foreground.
—Courtesy of California State Parks, Stout Research Center, © 2016

10. Petrified Forest, California

38° 33′ 19″ North, 122° 38′ 20″ West

The largest petrified trees in the world.

Defined as tall, woody plants distinguished from shrubs by greater height and a single trunk, rather than several stems, trees have cast both shadow and shade across the landscape since what seems like time immemorial. They function as haven for birds, squirrels, and a plethora of insects; add value to property; cleanse the air by absorbing pollutants; and retard soil erosion. Once they grew so thick in places the sun could not touch the ground, but today they cover only 30 percent of the land surface of Earth, a figure that has shrunk 40 percent since agriculture began 11,000 years ago.

The oldest known fossil specimens, dating back some 385 million years, evolved at a time when atmospheric carbon dioxide levels plummeted and global cooling gave rise to glaciation, a period of time leading up to the Devonian mass extinction during which as much as 70 to 80 percent of all animal species perished. During the later Carboniferous Period, the diversity and abundance of tree-sized vegetation expanded, leading to the creation of the vast deposits of coal that fueled the industrial revolution and today remain one of Earth's great reservoirs of stored energy. Trees have long been, and remain, a resource of unbounded importance.

Legislators in Arizona, Louisiana, North Dakota, Mississippi, Texas, and Washington have designated petrified wood as either the official state fossil or official state gemstone. Large numbers of petrified trees are found in at least twenty-two countries and more than ten states in America. In California, fossil tree enthusiasts are attracted to the Petrified Forest, a privately owned enterprise about 12 miles northeast of Santa Rosa and nestled in the rolling vineyard-laden hills of Napa Valley. First identified in a scientific publication authored in 1871 by Othniel Charles Marsh, the renowned paleontologist who described and named approximately five hundred species of prehistoric animals, this fossil enclave is today one of the world's finest examples of a Pliocene-age petrified forest.

Roughly 3.5 million years ago northern California was undergoing geological and biological transformation. Episodes of intense volcanic activity showered volumes of ash high in silica content upon the redwood giants of its forests. Entombed in an environment that prevented rot, the trees were slowly altered from fiber to rock. Groundwater enriched in silica leached from the volcanic ash permeated the wood and replaced the organic material cell by cell. This process was so exacting that today we can count annual growth rings and easily identify knots, burls, and bark texture. Several specimens in the Petrified Forest contain worm borings, and at least one an insect casing. Classified as the extinct redwood *Sequoia langsdorfii*, the aged stone replicas are all oriented in the same positions they held when they died. Examples along the self-guided 0.5-mile tour include the Giant, a petrified behemoth that measures 6 feet across, and the larger Queen, 65 feet long and 8 feet wide. After counting the Queen's tree rings, researchers determined she was some 2,000 years old when she was killed by the ashfall from a volcanic eruption. Needle and leaf remains of at least twenty other fossilized plant species have also been identified in this stony forest.

Permanently embalmed in minerals, the massive Pliocene inhabitants of the Petrified Forest remain a reminder that in the absence of trees Earth would indeed be a planet lacking in floral beauty.

Superbly preserved earth-tone tree rings characterized by contorted boundaries (left) and different thicknesses (right) are evidence that this 3.5-million-year-old fossil may have been subjected to life-threatening stress related to climate, fire, or disease. —Courtesy of the Petrified Forest

This mineralized 0.125-inch-diameter casing serves as proof that worms and trees have had a long history of coexistence.
—Courtesy of the Petrified Forest

Are these quartz-lined 0.125-inch-diameter holes evidence of two industrious wood-boring worms feeding side by side or just one that ate along a twisted and looped course? —Courtesy of the Petrified Forest

One has to wonder if this fossil knothole was ever home to a Pliocene-age squirrel or woodpecker.
—Courtesy of the Petrified Forest

11. La Brea Tar Pits, California

34° 03′ 47″ North, 118° 21′ 21″ West

An asphalt graveyard of ice age animals and plants.

The twenty-first century ushered in multiple crises, not the least of which was the realization that the era of cheap oil would end. Do the rocks of planet Earth contain a finite volume of this fossil fuel? Most certainly! Have the engines of progress consumed most of it? Certainly not! Since the age of oil began some 160 years ago, people have consumed approximately half of the world's reserves of conventional oil. However, this was the easy-to-find half. Discovering new reserves will increasingly require ever more costly and sophisticated exploratory technologies.

Early on, the process of discovering oil in the field was relatively simple: drill where there was an oil seep, the surface evidence of a pathway to a subterranean reservoir of petroleum. Edwin Drake sited his 1859 Pennsylvania borehole, the first well drilled specifically for oil in the United States, near seeps Native Americans had used to waterproof their baskets and canoes and mend their broken pottery. Between the 1860s and the early 1900s, geologists discovered new fields in California wherever they found an oil seep.

Of all the known seeps in the United States, one is unique: the La Brea Tar Pits in Hancock Park, along the 5800 block of Wilshire Boulevard in downtown Los Angeles. For almost 50,000 years naturally occurring asphalt, a petroleum by-product, has oozed at this 23-acre site from depths of 1,000 to 6,000 feet, slowly undergoing physical and biologic transformation, first by the evaporation of methane gas and then by chemical alteration by bacteria. Even today this gooey tar-black substance seeps onto the surface at an average rate of 10 gallons per day. The asphalt originates in mudstone and shale belonging to the Miocene-age Monterey Formation, a rock rich in diatom shells. Long ago the influences of temperature, pressure, and time changed the organic content of these microscopic, single-celled marine plants to petroleum.

During the hot summer months of the ice age, many an animal wandering at this site was entrapped within the sticky, viscous asphalt, their cries of panic quickly attracting numbers of predators and scavengers intent on feasting on an easy meal. Inevitably, additional opportunists would follow, and soon the site consisted of a large number of trapped carnivores, along with the first animal trapped, which according to the prevailing theory was most likely an herbivore. This scenario explains the documented ratio of seven carnivore skulls to every herbivore skull found at La Brea. After the flesh of these victims rotted, the pores of the bones became saturated with asphalt, a decay-inhibiting preservative used as an embalming agent since the days of the Egyptian pharaohs.

The result is one of the richest, best-preserved, and most intensively studied collections of Pleistocene-age fossils in the world. Paleontologists have recovered the remains of 60 species of mammals (including ground sloth, dire wolf, short-faced bear, Western camel, mammoth, mastodon, American lion, and saber-toothed cat), 135 species of birds (such as California condor, eagle, and falcon), and numerous plants and insects. Together, more than 620 species have been catalogued. The oldest La Brea fossil is 46,800 years old. After the dire wolf, the most common is the saber-toothed cat, the infamous and extinct 6-foot-long, 750-pound pouncing predator that is California's state fossil. The remains of only one human have been found: the 9,000-year-old skull and partial skeleton of La Brea Woman.

Carnivore fossils (wolf, saber-toothed cat, coyote, lion) outnumber herbivores seven to one at La Brea, chilling evidence that it was a popular killing ground.

Iridescent bubbles of methane gas percolating in a sea of petroleum froth show chemical activity is alive and well at La Brea. —Courtesy of Sarah Lansing

 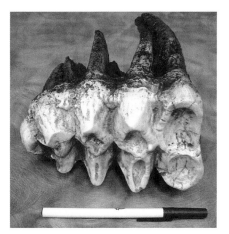

The molars of mammoths and mastodons, both herbivores, are radically different: those of mastodons (right) have high crowns and are well suited for mashing leaves and tree branches, whereas those of mammoths (left) resemble a cheese grater and are suited to grinding up tough grasses. Pens for scale.

12. Sharktooth Hill, California

35° 26′ 47″ North, 118° 53′ 55″ West

Tropical-oasis fossils entombed in a desert.

A complete analysis of the earthen layers exposed in the wall of almost any vertical shaft dug into the arid countryside of Kern County, the southernmost county in the 450-mile-long Central Valley, might well contain information so voluminous that it could not conveniently be contained in one book. Several volumes would be required. The first, with an emphasis on the soil horizon, would highlight that the county is lauded as a cornucopia so overflowing with food that it is debatable whether it should be called the "bread basket" or the "salad bowl" of America. Dubbed the most diverse and productive farming region in the world, the desert terrain produces more than 240 varieties of crops, supplying lunch boxes, restaurants, and dinner tables with a diverse range of eatables, including grapes, beef, pistachios, and potatoes.

Digging deeper, volume two would probe the secrets of the 6-to-12-million-year-old Monterey Formation, an energy-rich rock that guaranteed Kern County a prominent position in the history of oil production. When black gold bubbled to the surface of a hand-dug well in 1899 in the farm village of Bakers Swamp, an oil boom was initiated and a forest of wooden derricks grew into a city the leading citizens wisely renamed Bakersfield. More than forty thousand wells dot the county, delivering 70 percent of the conventional oil produced in the Golden State, close to 10 percent of that produced in the United States, and almost 1 percent of that produced in the world.

The third volume would investigate a resource of an entirely different nature, created at different depths and of greater interest to a paleontologist than a farmer or a petroleum geologist. Inundated by a shallow sea 15 million years ago, the entire southern Central Valley abounded with a heterogeneous population of marine organisms. The tectonic pulse of Earth was governed by tranquility, and life was good, but not without irritation. Periodically, flood-stage rivers flowing off the western slopes of the Sierra Nevada transported great volumes of silt and mud into this tropical oasis, killing organisms and preserving them as fossils. Subsequent events compressed the sediment into rock of the Temblor Formation and uplifted it as Sharktooth Hill, an exquisite fossil bed in northeast Bakersfield.

Legendary Swiss geologist Louis Agassiz first documented the exceptional quantity and quality of this hill's fossils in 1856. Since then the site has yielded ten types of dolphin, five varieties of whale, four extinct turtles, twenty species of bird, and two nearly complete 10-foot-long skeletons of *Allodesmus* (a precursor to the modern sea lion), as well as the fossilized remains of a sea cow, porpoise, walrus, seal, and marine crocodile. Teeth of more than two dozen species of shark and ray gave namesake identity to the site. Sharktooth Hill fossils are concentrated in a narrow 1-to-4-foot-thick layer of the middle Miocene Temblor Formation.

The casual visitor may view the Bakersfield area as one focused on the economies of nourishment and crude oil. Visitors attuned to interests of deep time and biologic bounty, however, will find a treasury of ancient life acclaimed to contain the world's greatest density of middle Miocene marine vertebrate fossils. A shovel, mallet, and flat-headed screwdriver are the recommended means of accessing the wonders of Sharktooth Hill, at the Ernst Quarries, on Round Mountain Road, open to the public on a fee basis.

Four partially exposed vertebrae of a juvenile baleen whale from Sharktooth Hill. This fossil rock specimen is about 30 inches wide, left to right. —Courtesy of Rob Ernst

Fifteen-million-year-old teeth of the cow shark, thought to be the most primitive of all sharks, which fed on crustaceans, other sharks, and the occasional marine mammal. Dating back to the Cretaceous Period, cow shark fossils are found throughout the world. Photo is 2.75 inches across. —Courtesy of John Cartier

Sharktooth Hill specimens of Isurus hastalis, *the extinct ancestor of the contemporary mako and white sharks, surround a tooth of* Megalodon *(center), one of the largest and most potent predators of vertebrate history.* —Courtesy of Rob Ernst

Tooth of Desmostylus, *an extinct, 6-foot-long, herbivorous, hippopotamus-like mammal that lived in estuaries and fed on freshwater plants.* —Courtesy of Rob Ernst

13. Marsh–Felch Quarry, Colorado

38° 32′ 06″ North, 105° 13′ 18″ West

Lesser-known nineteenth-century battles fought on the Rocky Mountain plains.

Throughout its history, America has fought many wars, four of which historians classify as major, a status determined by measures of intensity and significance. World War I and World War II were conducted at foreign venues in an attempt to protect the world from political disorder and the rise of dictatorships. Two homegrown episodes—the Revolutionary War and the Civil War—were fought to give birth to a nation dedicated to independence and then to preserve that nation when it teetered on the edge of self-destruction. A fifth and lesser-known conflict, really a series of skirmishes spread across the Rocky Mountains, was a war of personalities, wits, and deception. The "bone wars" were conducted by two prominent paleontologists, men of distinction driven by a policy of do or die. They developed an antagonistic relationship that gained notoriety and ended only with death.

When rumor became fact, in the winter of 1876–1877, that novice naturalists had discovered a treasure pile of vertebrate bones in a canyon north of Cañon City, Edward Drinker Cope of the Academy of Natural Sciences in Philadelphia was the first paleontologist of authority to react. Learning that Oramel Lucas, one of the novice fossil hunters associated with the discovery, had assembled five wagonloads of fossils, Cope made arrangements for Lucas to begin excavations from Jurassic-age Dakota Group strata near Cañon City. Because the rock matrix was soft and easily worked with hand tools, a number of important specimens were soon unearthed and sent to Philadelphia. On August 23, 1877, Cope announced the discovery of *Camarasaurus supremus*, a 150-million-year-old long-necked plant-eating dinosaur. And with that, the opening salvo of the bone wars echoed through the halls of academia.

Notified that Cope was making reputation-enhancing news with his Colorado discoveries, Othniel Charles Marsh of the Academy of Natural Sciences of Yale University lost little time making arrangements to excavate a second site north of Cañon City, in association with Marshall P. Felch, a local rancher. Operations were underway at the Marsh-Felch Quarry in the final months of 1877. Thirty-five boxes of dinosaur bones were soon wrapped and crated for protection, transported in horse-drawn wagons to Cañon City, and shipped by rail to New Haven, Connecticut, where Marsh examined them. Driven by a desire to surpass Cope's degree of success, Marsh quickly announced the discovery of two new genera of dinosaur: *Allosaurus*, a member of the subgroup that includes all flesh-eating dinosaurs, and *Diplodocus*, a long-necked, long-tailed, plant-eating colossus that approached 100 feet in length.

In the years that followed, both fossil-supremacy-seeking contenders continued excavating their respective sites, all the while engaging in nefarious activities designed solely to disgrace and distract each other. In time both men were financially and socially ruined. Cope died in 1897, and Marsh was laid to rest two years later.

Cope's quarry is not open to the public, but the Marsh-Felch Quarry, 6.5 miles north of Cañon City on the west side of Garden Park Road, was designated a National Natural Landmark in 1973. There, interpretive displays set the stage to such an extent that one can almost hear the echoes of shovels and chisels chipping away at yet another fossil specimen. The bone wars are now history, but that history records a time when the science of paleontology was young and a few paleontologists were eager to use any means possible to garner academic kudos and contribute to the history of vertebrate evolution.

Bones of sixty-five individual dinosaurs were extracted from the rocks overlying the gray cliff (center) at the Marsh-Felch Quarry.

Many of the dinosaur bone fragments extracted from the Cope and Marsh quarries were of poor quality, leading to incorrectly interpreted complete skeletons. Pen for scale. —Specimen from the collection of Steven Jones

A sepia-tone stereograph showing the 1877 excavation site where Oramel Lucas unearthed Camarasaurus supremus. —Courtesy of the Royal Gorge Regional Museum and History Center

14. Picket Wire Track Site, Colorado

37° 39' 35" North, 103° 34' 18" West

The largest dinosaur track site in North America.

The nature of the climate of the plains of southeastern Colorado can be readily described with a handful of characteristics: low relative humidity, infrequent rain and snowfall, extreme daily and seasonal temperature changes, and abundant sunshine. The snows are commonly wind driven, and the rains are often charged with some of the continent's most ferocious hail. Finally, and perhaps most significantly, although nourished by an average of 16 inches of precipitation per year and drained by the 200-mile-long Purgatoire River, the region is almost always in or on the verge of drought.

About 150 million years ago, the setting was quite different. Pangaea, the great supercontinent that had formed during the late Paleozoic Era, had reached old age—geologically speaking—and was beginning to fracture. North America was in the early stages of development, but many millions of years would pass before the Rocky Mountains began their rise to dominance. Southeastern Colorado enjoyed a tropical climate and was inundated by Dinosaur Lake, a shallow body of freshwater teeming with colonies of clams and schools of fish. Onshore, forests of cone-bearing trees towered over an understory of verdant ferns.

Today the ancient shoreline of Dinosaur Lake is identified by ripple marks, mud cracks, and, as the name suggests, fossil footprints made by generations of migrating dinosaurs. The best known of these shoreline locales is the Picket Wire Track Site, a dinosaur freeway preserved in rocks of the Jurassic-age Morrison Formation. More than 1,300 footprints compose the 100 separate trackways that extend along a 0.25-mile-long section of the Purgatoire River in Las Animas County. Numerous fossil bones have been discovered in the nearby canyon walls, but the trackways contain only the prints of *Allosaurus* and *Brontosaurus*, two very different dinosaurs.

Brontosaurus, the "thunder lizard" that has captured the imagination of generations of children, stood 15 feet high, was some 70 feet long, and weighed more than 30 tons. In spite of its imposing form, the brontosaur was a gentle four-footed giant with a long neck and an even longer whiplike tail. Because its diet was restricted to plants and leaves, paleontologists believe it had to eat on a nearly twenty-four-hour basis in order to satisfy its appetite. Characterized by massive, saucer-shaped indentations, brontosaur footprints comprise 40 percent of those found at Picket Wire.

The remaining prints are those of *Allosaurus*, a dinosaur with a ferocious and unforgiving character. A stocky, 2.5-ton, 30-foot-long, 16-foot-tall, bipedal meat-eater equipped with a massive head and jaws lined with dozens of sharp, serrated teeth, it was the alpha predator of the Jurassic Period. It preferred to snack on herbivores. Because it possessed a rather weak bite, *Allosaurus* slammed its teeth into its victims and tore them apart. Sharp claw marks accentuate its three-toed footprint. Several Picket Wire trackways show that brontosaurs walked side by side, evidence that perhaps these massive cowlike animals lived in herds. In contrast, tracks of the meat-eating *Allosaurus* radiate in all directions, indicating they stalked the lakeshore as solitary predators.

Located 25 miles south of La Junta and accessible by an 11-mile round-trip hike that begins at Withers Canyon Trailhead, the Picket Wire Track Site is a chapter in paleontological history in which the contest between carnivore and herbivore is written in stone.

Location of the Picket Wire Track Site.

A 150-million-year-old trackway left by many brontosaurs walking side by side. Erosion by the Purgatoire River has exposed this trackway, but during flood stage the same river is gradually destroying the tracks. —Courtesy of the US Forest Service

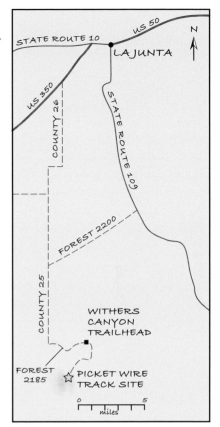

The Picket Wire Track Site along the Purgatoire River, as viewed from the air. At least ten trackways can be seen, some made by solitary animals and others by several dinosaurs strolling together.
—Courtesy of the US Forest Service

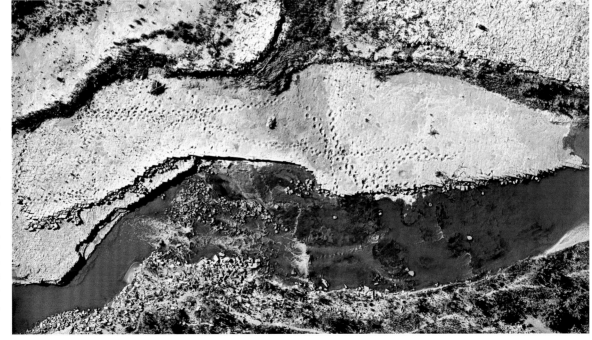

15. Denver Museum of Nature and Science, Colorado

39° 44′ 55″ North, 104° 56′ 34″ West

A prehistoric journey that's as easy as putting one foot in front of the other.

The United States doesn't lack for museums with the mission of exploring Earth's amazing story of life by way of fossil displays and related landscape dioramas. Some museums emphasize local paleontology, whereas others focus on regional scenarios; some utilize static displays, and others delve into the intricacies of animation. For good reason, only a few attempt to explore the evolution of life on the grandest of all scales, from the very beginning of Earth's existence to the events of yesterday.

It's easy to state that Earth is 4.6 billion years old, but comprehending such a statement is something altogether different. Even individuals involved, on a daily basis, with manipulating large numbers have little understanding of the term *billion*, let alone *4.6 billion*. A little verbal exercise might help clear away the fog: if you began counting right now at a normal rate—one, two, three, four, and so on—you would reach one billion (1,000,000,000) in approximately thirty-two years. Or, consider this: one billion US dollar bills taped end to end would circle Earth four times, a distance of nearly 100,000 miles.

Located in the heart of the Mile High City at 2001 Colorado Boulevard, the Denver Museum of Nature and Science is one of the few American museums that has attempted to document the deep-time passage of life on Earth since its inception. Its exhibits seek to educate the public in every aspect of nature and science. The paleontology sector, titled "Prehistoric Journey," traces the evolution of life on Earth, from the development of single-celled organisms in the very earliest of days through the reign of dinosaurs during the Mesozoic Era to the varieties of life that characterized the glacial world of the Pleistocene Epoch, and ends with the inhabitants of today's world.

Around every corner and along each wall the story unfolds in chronologic order, portraying yet another episode in the continuing tragedies and triumphs associated with the evolution of Earth's flora and fauna:

- Imagine standing on the shoreline of a vast ocean covering Kansas during the Permian Period, when a *Ctenacanthus* shark prowled the offshore depths and *Meganeura* dragonflies soared high utilizing 2-foot wingspans.

- Let your imagine run wild, but don't stand too close to the exhibit of a family of heavily armored, herbivorous *Stegosaurus* under attack by a ravenous, carnivorous *Allosaurus*, one of the earliest Jurassic dinosaurs discovered.

- Marvel at the 4-foot length of *Platyceramus platinus*, an extinct Cretaceous clam that bears the distinction of being the largest bivalve mollusk.

- Peer into the toothless mouth of *Dunkleosteus terrelli*, an armored terror of the Devonian seas that grew to lengths as long as three automobiles lying end to end and employed the sharp edges of its jaws to slice up fellow fish.

- Take in *Eusthenopteron*, a lobe-finned saltwater creature that used primitive limbs to explore land and expand its diet. This fish led the way for future generations to evolve as amphibians.

- Contemplate the chaotic arrangement of fossil body parts encased in a small portion of rock from a much larger bone bed made up of hundreds of skeletons of *Menoceras*, one of the smallest rhinoceroses to roam the wilds of North America.

Open daily and year-round, except on Christmas Day, this prehistoric journey is a defining exhibit of interest to novice and professional paleontologists alike.

This 16-square-foot section of Ordovician-age seafloor limestone contains the remains of a dozen crinoids. The two specimens to the right contain all the principal body parts: arms, calyx, stem, and root system.
—© Bailey Library and Archives, Denver Museum of Nature and Science

Discovered in the Jbel Wawrmast Formation of Morocco, this 530-million-year-old, 18-inch-long Acadoparadoxides briareus, *ancestor to the modern-day horseshoe crab, was one of earliest and largest trilobites.* —© Bailey Library and Archives, Denver Museum of Nature and Science

This 5.4-foot-high palm frond from the Green River Formation of Wyoming is prime evidence that as recently as 40 million years ago the Cowboy State was much like Florida is today. —© Bailey Library and Archives, Denver Museum of Nature and Science

A 3.5-square-foot section of the Lias Formation of England encasing ammonites. Does this fossil slab represent mass death related to climate change or a catastrophic event, or does it encapsulate a locale where creatures went to die? —© Bailey Library and Archives, Denver Museum of Nature and Science

16. Peabody Museum of Natural History, Connecticut

41° 18' 57" North, 72° 55" 16" West

Eastern personalities and western boneyards that formed the basis of today's dinomania.

In the minds of many historically minded paleontologists, any reference to the Peabody Museum of Natural History is license for an impromptu mention of Othniel Charles Marsh. Born in 1832 into a struggling farm family in Lockport, New York, Marsh showed early interest in collecting fossils. Educated through the financial support of his wealthy uncle, philanthropist George Peabody, Marsh persuaded Peabody to include Yale University in his choice of beneficiaries. The Peabody Museum of Natural History was founded in 1866 with an endowment of $150,000—today equal to $2.2 million—and Marsh was named Yale professor of paleontology, the first appointment of its kind in the United States.

Characterized as possessive, egotistical, and occasionally unscrupulous, today Marsh is often lauded as the most important American paleontologist of the nineteenth century. Specimens from his vast and celebrated collection of mammal, bird, and dinosaur fossils fill the two-story Great Hall of Dinosaurs, the preeminent display of the Peabody Museum of Natural History, which houses one of the largest and most extensive fossil collections in the United States.

When informed of the discovery of numerous dinosaur fossils in the barren landscapes of Colorado and Wyoming, Marsh organized and led the first of four Yale-sponsored expeditions west in 1870. At the time, the American inventory of dinosaur genera numbered a mere thirteen. Within a decade his hired hands had filled hundreds of boxes with fossils from copious boneyards in Cañon City and Morrison, Colorado—the latter the site of discovery for *Stegosaurus*, *Allosaurus*, and *Apatosaurus* (formerly known as *Brontosaurus*)—and Como Bluff, Wyoming.

Fueled by fire-in-the-belly ambition, Marsh hoped to possess a world-class fossil collection, but Edward Drinker Cope, a fellow paleontologist and distinguished comparative anatomist, offered significant competition. Their onetime close friendship was damaged beyond repair when Marsh pointed out that Cope, while reconstructing the marine reptile *Elasmosaurus*, had erringly placed the skull on the tip of its tail, rather than at the top of its neck.

After that, the pursuit of fame, name, and reputation became a full-fledged, nasty rivalry known as the "bone wars," a nearly three-decade-long period of bribery, trickery, theft, destruction, and spying. It only ended with the death of Cope, in 1897, and Marsh, two years later (see also site 13). This nefarious and contentious chapter of paleontology history is, however, not without merits. The bone wars resulted in the identification of more than 130 megafauna genera and species of the Mesozoic Era, nearly 60 percent of them attributed to Marsh.

By the time of his death, Marsh had gained the title of America's most renowned paleontologist and praise from the originator of the theory of evolution. In a letter penned to "My dear Prof. Marsh," dated August 31, 1890, Charles Darwin wrote, "Your work . . . on the many fossil animals of N. America has afforded the best support to the theory of evolution, which has appeared within the last 20 years . . . and I can say nothing stronger than this" (McCarren 1993).

Knowledge of the ego-driven rivalry between Marsh and Cope has diminished with time, but it is not forgotten. The vast catalogue of information they forged through competition and conflict helped build the foundations of dinomania, the full-flush interest—both amateur and professional—in the Age of Dinosaurs. Fossils on display in the Peabody Museum of Natural History give three-dimensional definition to the concept of dinomania.

Typically, an Albertosaurus *skull contained as many as fifty-eight razor-sharp teeth and grew to more than 3 feet long. This massive specimen is characterized by large openings called fenestrae, the triangular voids in the center that provided attachment points for muscles and sensory organs.* —Specimen from the Peabody Museum of Natural History collection

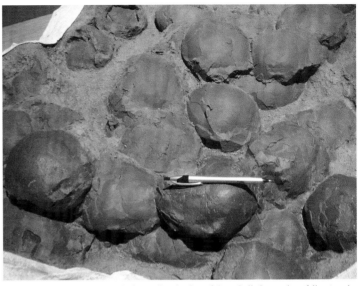

Titanosaurus *nests consisted of irregular clutches of thin-shelled eggs that, following the pattern of her species, the mother abandoned. The 2-foot-long hatchlings grew to exceed 75 tons, making this herbivore one of the heaviest animals to ever walk on Earth. Pen for scale.* —Specimen from the Peabody Museum of Natural History collection

Perhaps this close-knit herd of Psittacosaursus *juveniles, intertwined with one adult, formed an impromptu coalition seeking a sense of security—a 120-million-year-old example of safety in numbers. The adult skull at left center is 9 inches wide.* —Specimen from the Peabody Museum of Natural History collection

17. Chesapeake and Delaware Canal, Delaware

39° 33′ 42″ North, 75° 34′ 39″ West

The 7-inch-long forebearer of the octopus and squid.

The chronicle of evolution that has occurred in the Phanerozoic Eon can be described simply as long periods of progress, measured by increases in faunal diversity and abundance, interrupted by history-churning episodes of extinction. Regarding the latter, studies focus on the "big five," those mass extinction events that define the end of the Ordovician, Devonian, Permian, Triassic, and Cretaceous Periods of geologic time. The Permian occurrence was the most serious, involving the eradication of up to 96 percent of all marine species, but the Cretaceous event is the one that creates academic buzz and captures the headlines.

Numerous theories have been proposed for the Cretaceous episode of death: pandemic disease, climate change, magnetic field reversal, sea-level change, extraterrestrial radiation, and the alteration of Earth's surface through plate tectonics. These are noteworthy, but scientific consensus centers on two massive catastrophic events that occurred 66 million years ago: the impact of the Chicxulub meteorite on the Yucatán Peninsula of Mexico, and the eruption of 123,000 cubic miles of lava that formed the Deccan Plateau of west-central India. One or both of these events likely precipitated the extinction event (see also site 3).

While the cause of the extinction continues to be debated, the effects are well known: as much as 85 percent of the world's species succumbed to the everlasting sting of extinction. In the company of paleontologists, much ado is made of the demise of nonavian dinosaurs, flying and swimming reptiles, and numerous species of mollusks, marine plankton, and flowering plants. However, the roll call of Cretaceous extinction victims also includes the belemnite, the little-known ancestor of the modern-day squid and octopus.

Belemnite fossils have long been a matter of curiosity. Early Egyptians associated them with male fertility because of their phallus-like shape, and English peasants once pulverized them and blew the dust into the eyes of their horses, a supposed remedy for infection. Medieval Scandinavians considered them the candles of forest elves, and their feudal neighbors in Germany were in basic agreement, naming them ghostly candles.

In July of 1996 *Belemnitella americana* was designated Delaware's official state fossil. Reminiscent of its modern-day squid cousin, this Late Cretaceous belemnite, like all belemnites, had an internal shell covered by a leatherlike skin, a cluster of tentacles studded with hooks useful for grasping prey, and a water-expelling siphon system that provided jet propulsion.

The spoil piles along the north side of the Chesapeake and Delaware Canal, immediately east of the Reedy Point Bridge in Delaware City, are a great place to find belemnite fossils. With a little patience, one can fill a ziplock bag with calcareous, pencil-shaped, amber to orange belemnite *guards*—the rear-end portion of the internal shell that is commonly fossilized—especially if the US Army Corps of Engineers has recently dredged the adjacent canal. The sole purpose of the guard was to counterbalance the weight of the animal's head and arms. Without it, belemnites would have tipped forward, head down, and been unable to either swim or feed.

Belemnites were a favorite food of the critters that occupied the top of the Cretaceous food chain. One European shark fossil's stomach cavity was filled with 250 guards, possibly the singular fossilized example of a shark dying of a stomachache after bingeing on belemnites. Fossil guard remains at the Chesapeake and Delaware Canal are found with reptile vertebrae and fish teeth, as well as palm-sized specimens of extinct pelecypods.

Ranging from 1 to 4 inches in length, the fossils in greatest abundance at the Chesapeake and Delaware Canal site are belemnite guards, the tubelike, amber-brown structures that formed the internal shells of the living animal.
—Courtesy of Jayson Kowinsky

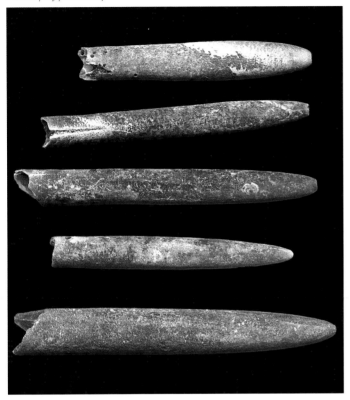

The oyster Exogyra costata, *common in the spoil piles near Reedy Point Bridge, is an index fossil for the Cretaceous Period. Ruler for scale.* —Courtesy of Jayson Kowinsky

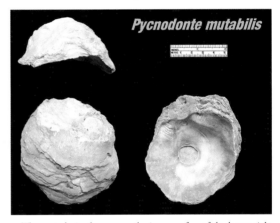

The central, circular scar on the inner surface of the lower-right oyster shell marks the position of Pycnodonte mutabilis*'s adductor muscle, which it used to close its two valves. Fossils of this species are also common in the spoil piles. Ruler for scale.*
—Courtesy of Jayson Kowinsky

The spoil piles near the base of the Reedy Point Bridge are considered the best fossil site in Delaware. Small-scale collecting for personal collections is permitted.

18. Bone Valley, Florida

27° 13' 55" North, 81° 53' 29" West

Shark teeth and mammoth molars from the river of tranquility.

Identified on sixteenth-century Spanish charts as Rio de la Paz, and named Talakchopcohatchee by the Seminole Tribe, the 106-mile-long Peace River dominates the lowland terrain of southwestern Florida. Born in the marshlands of Polk County, it meanders south, weaving east of Bartow and west of Bowling Green and through Zolfo Springs and Arcadia before emptying into the Gulf of Mexico at Punta Gorda. Deposits of phosphate, an essential component of fertilizer, have been mined from the river basin since the 1880s, but the plethora of fossils that erode from Miocene-to-Pleistocene-age sedimentary bedrock is the resource that gives the Peace River an understandingly high "wow factor" reputation.

The fossils of the Peace River represent a true supermarket collection of mammalian and marine fauna that lived between 16 million and 10,000 years ago. Large and small, complex and simple, they represent a varied assemblage of Cenozoic life: bison, llama, ground sloth, horse, beaver, tapir, dolphin, gator, whale, barracuda, mastodon, mammoth, and shark (including the teeth of some eight to ten species).

The fossilization of any organic material is, even under ideal conditions, a hit-and-miss proposition—mostly miss. The best opportunities are created by burial in fine-grained sediment in an environment lacking oxygen and bacteria. Flesh, soft internal organs, cartilage, and muscle tissue decompose first, followed by bone and shell. Because teeth have a core of pulp packed with nerves and blood vessels, are coated with dentine (a form of calcified tissue), and are crowned with a protective layer of enamel, they commonly resist the processes of decay.

Toothlike structures first appeared almost 500 million years ago in jawless fish in the form of biting plates. As tetrapods, reptiles, and mammals evolved, teeth shape, size, and number and mouth position changed. Today, four types of mammal teeth are recognized: incisors for grasping and ingesting nutrients; daggerlike canines for stabbing and biting; premolars for slicing and dicing; and molars for crushing and grinding. The crown jewels in the Peace River catalogue are mastodon and mammoth molars measuring up to 9 inches long, 5 inches wide, and 9 inches tall.

Canoe Outpost, 2.3 miles west of Arcadia and off Florida 70, offers public access to the Peace River's tooth treasures. Success is dependent on a few simple rules: Collect only from river sediment, remember that riverbanks are private property, concentrate on eroded shorelines and gravel banks within the bends of the river, and avoid well-visited stretches near boat ramps. Protected by sunscreen and hip boots, get out of the canoe and into the water, stir the bottom with a paddle or a metal rod, and shovel the gravel into a spaghetti strainer or sifter screen. Wash away mud and sand, examine what remains with a practiced eye, and take home the best specimens. It helps to collect on sunny days and when the water level drops at least 12 inches below normal, which happens most often during March, April, and May.

Local residents use the fanciful nickname "bone valley" for the Peace River, a justly deserved name. People regularly find alligator, turtle, shark, whale, and dolphin specimens, as well as fragments of dugong ribs and whale and dolphin ear bones up to 6 inches in length. During the thousands-of-years process of turning bone to stone, fossils often absorb chemicals from the enclosing sediment and turn black, making them easy targets for both novice and expert. In bone valley, the key to fossil-finding success is to get wet, look for black, and not give up.

The majority of bone valley fossils are from marine animals, but on occasion people locate molars of ice age mammoths. Pen for scale.

Once they are buried, shark teeth, such as this museum-quality 5.5-inch-long Megalodon *specimen, absorb minerals from the surrounding sediment and turn shades of dark brown, black, or gray. Pen for scale.*

The fossil fragments of dugongs, a marine mammal, are a relatively easy find in bone valley because of their black color. Penny for scale.

A collection of Peace River keepers: miscellaneous shark teeth, a prized 3.6-inch-long Megalodon *tooth, and a fragment of turtle shell (bottom). Penny for scale.*

19. Florida Caverns State Park, Florida

30° 48′ 31″ North, 85° 12′ 47″ West

Fossils large and small, exotic and common, preserved in a subterranean vault.

The Sunshine State is a land grounded on bedrock of limestone, a soft, light-colored sedimentary rock composed of calcium carbonate. Formed from the skeletal remains of small organisms, Florida limestone is used to make Portland cement, is cut into slabs of architectural stone, and is even fed to chickens so they produce stronger eggshells. In its natural in situ state, however, it is the bane of anyone interested in maintaining the natural contours of the landscape, because it is very susceptible to the development of karst.

Karst is a topography characterized by caves, sinkholes, and subsurface drainage, all of which form through dissolution. In certain Florida locales, karst gives the countryside a Swiss cheese–like complexion. Some of the most visible examples are found in Jackson County, home to Florida Caverns State Park (north of Marianna), the only air-filled cave in the state open to the public. The hour-long tour at the park allows one to not only observe a prime example of convoluted subsurface karst, but also to discover fossils unusual in both shape and design.

Native American hunters frequented the Florida Caverns more than 1,000 years ago, as evidenced by artifacts and footprints. Accentuated by a dazzling display of stalagmites, stalactites, soda straws, and flowstone draperies, the mile-long system of passageways winds through porous rock deposited in a tropical sea 40 to 25 million years ago, during the Eocene and Oligocene Epochs. Fossils are distributed along the cave walls and ceiling, though the stained surface can obscure them, a situation easily remedied with flashlight illumination. Common Florida Panhandle specimens of coral, bryozoans, and brachiopods are present, but the megasized representatives of nautilus, echinoid, and foraminifera are the ones you've gotta see.

Foraminifers, or forams, are primarily marine organisms that live and feed in oceanic sediments. They are so numerous, it is estimated that approximately 35 percent of the world's oceans are blanketed with ooze composed of their chambered, skeletal remains, which are the size of sand grains. Because some species are found in specific environments and others are geologically short-lived, they are useful for understanding climate change, ancient environments, and where new reserves of petroleum and natural gas might be discovered. Several genera of unusual, gargantuan foraminifera are found in the caverns' strata, the most conspicuous being the 2-inch-wide, multirayed *Asterocyclina*, and the wafer-shaped, nickel-to-quarter-sized *Lepidocyclina*. The genera of these giant specimens are closely related to fossils in the building stones of the great pyramids of Egypt.

About forty walnut-sized species of echinoid, otherwise known as sea biscuits and sea urchins, are also present. With appearances suggesting an alien genesis, their skeletons are enveloped by interconnected plates with a fivefold symmetry. These plates were once covered with radiating spines designed to protect against predators.

The title for prize fossil of the caverns, however, belongs to the extinct, softball-sized pearly nautilus *Aturia*. It cruised the Eocene ocean with twenty-first-century technology: a series of chambers that allowed it to maintain submarine-type buoyancy and move with jet propulsion by fluctuating the amount of water in its various chambers. *Aturia* satisfied its appetite with a combined strategy of scavenging and predation.

Flowstone can mask the fossils in the cave rocks. However, there are multitudes of *Lepidocyclina* in the building stones of the visitor center. Inside or outside, in sunshine or subterranean obscurity, a Florida Caverns tour is an adventure into a wonderland defined by fossils and geologic time.

The uncommonly large, starlike foraminifera Asterocyclina *weathers loose from the Florida Caverns limestone. The large one in the middle is 3 centimeters wide; the small ones are 1 centimeter wide.* —Courtesy of Harley Means, Florida Geological Survey

A small saucer-sized Clypeaster *sea urchin illuminated by a flashlight. Note its faint but characteristic fivefold-symmetry radiating outward from the animal's mouth (center) and the location of its anus (eleven o'clock position).* —Courtesy of Harley Means, Florida Geological Survey

The fossil remains of the extinct nautilus Aturia, *which trolled open waters in search of the number one food item on its menu—small fish.* Aturia *is the size of a volleyball.* —Courtesy of Harley Means, Florida Geological Survey

20. Windley Key Fossil Reef Geological State Park, Florida

24° 57′ 02″ North, 80° 35′ 44″ West

A twentieth-century tale involving the Overseas Railroad and an ice age coral reef.

Henry Flagler, a name synonymous with modern-day Florida, was possessed with a dream. By 1896, his Florida East Coast Railway reached from Jacksonville south to the unincorporated village of Miami, a settlement of fewer than fifty people. His vision, however, stretched even further: Why not extend the railbed another 165 miles, all the way across the Florida Keys? On January 22, 1912, blind but still very much in charge of his affairs, he inaugurated regular service on the Overseas Railroad to its termination in Key West.

Originally he had hoped to build the extension across the southern tip of Florida to Cape Sable, where a ferry service would connect it to Key West. When the miseries of the Everglades proved insurmountable, however, the Keys route (today US 1), with its ready access to Key Largo Limestone bedrock, was approved. Quarries supplied thousands of tons of rock to construct the railbed and the many bridge approaches. After the railroad was completed, the Windley Key quarry continued to operate until the 1960s, supplying the decorative rock called "Keystone" to the building industry. Today, Windley Key Fossil Reef Geological State Park, at mile marker 85.5 near Islamorada, is a treasure for aficionados of railroad history and anyone interested in comparing contemporary corals with their fossilized ancestors.

Key Largo Limestone is a cream to light-gray, highly porous and permeable sedimentary rock composed of entangled thickets of fossil coral heads encased in a matrix of calcite. The term *coral* refers to bottom-dwelling invertebrate marine organisms that live in an outer skeleton constructed of calcium carbonate. Confined to locales characterized by sunlight and relatively shallow water, colonial forms develop growths as dense as those found in any impenetrable jungle.

The Florida Keys chain of islands was in full coral bloom 125,000 years ago, but then Earth began to cool, continental glaciers extended their frigid footprints, sea levels dropped more than 300 feet, and reefs were eroded. Eventually the widespread lobes of ice began to melt, sea levels rose, the reefs flooded, and a new cycle of coral growth began. This cycle continues today in the shallow waters offshore Windley Key Fossil Reef Geological State Park, fueled by sea-level rise associated with global warming.

The centerpiece of the park is the half-acre quarry floored by Key Largo Limestone. A complex framework of Pleistocene-age corals, preserved in their original moment-of-death growth positions, are embedded in the quarry's 8-foot-high walls. Close inspection occasionally reveals 2-inch-long imprints of *Lithophaga*, the stone-eating mollusk, but the fossil mass is overwhelmingly composed of two frame-building genera of corals: *Montastraea annularis* and *Diploria* (brain coral). Cross-sectional views of the former display a starburst pattern of skeletal elements projecting inward, whereas brain coral resemble the human brain.

Today, reef development is being affected by bleaching, an alteration in color related to human-induced warming of the world's oceans. Corals subjected to unusually warm water expel the algae living in their tissues, causing the coral to turn white, a condition that may lead to death. This environmental change makes the fossil reefs of Windley Key all the more significant to students of paleontology. They serve as relics of a time when the activities of Stone Age communities were conducted in relative harmony with Mother Nature, unlike the relationship seen with humans and the environment today.

Cross section of a cluster of Montastraea annularis, *also known as boulder star coral. Found throughout the Caribbean, this species is on the endangered list, a victim of bleaching. Quarter for scale.* —Courtesy of Rachael Garrity

Lithophaga, *the giant date mussel, bores into limestone and fossil coral rock with the help of acidic secretions. This club-shaped boring is actually a trace fossil.*

Close-up of Diploria. *Modern-day examples of this brain coral grow upward approximately 0.15 inch per year, achieving domelike shapes 5 to 6 feet in diameter. This coral feeds on zooplankton and bacteria. Quarter for scale.* —Courtesy of Windley Key Fossil Reef Geological State Park

21. Brevard Museum of History and Natural Science, Florida

28° 23′ 14″ North, 80° 45′ 48″ West

The story of Florida's earliest-known snowbirds.

Before the end of the nineteenth century, anyone considering travel along the Atlantic coast of Florida had to be endowed with a large dose of pioneering spirit. The land was then a composite of swamp, mosquitoes, heat, and humidity. When Henry Flagler, cofounder of the Standard Oil Company, arrived in St. Augustine in the early 1880s, he found the environment agreeable but lacking in visitor amenities. With the vision that makes rich men richer, he embarked on a program of construction that by 1912 had extended rail and hotel facilities across the Florida Keys to Key West. Flagler is lauded today as the man who, economically speaking, put the sun into the Sunshine State, but a discovery made seven decades after his death showed that pioneering visionaries of a much earlier era had preceded him.

It is not known where these inhabitants came from, but extensive analyses have established that a population of very early Americans lived in the Titusville area more than 7,000 years ago, some 2,000 years before the Egyptians built their pyramids and thousands of years before most of the Iron Age bog bodies of Europe were laid to rest. In 1982, during the excavation of Windover Bog, a small peat-filled marsh, the remains of more than 168 individuals were found entombed in an oxygen-free, low-acidity, submerged cemetery. The find is acclaimed as one of the world's greatest discoveries of ancestral humans.

Because many individuals and their accoutrements have been recovered from the site, researchers have been able to prepare a demographic profile of the Windover population. The women were on average 2 inches over 5 feet tall, and the men 4 inches taller. Half of the skeletons are those of children. Bone structure indicates the adults were husky and well muscled, and many lived to a surprisingly ripe old age of seventy years. Pain and accident, however, were a part of daily life: one sixteen-year-old lad suffered from the crippling effects of spina bifida, and a fifty-year-old woman sustained multiple bone fractures prior to her death.

Life in the Windover village revolved around survival priorities, the most important being hunting and gathering food. Nourishment came from a variety of sources, including deer, rabbit, fish, snails, manatees, nuts, cattail roots, and grass and berry seeds. The stomach of one woman contained more than three thousand elderberry seeds, possibly the remains of a deathbed dessert.

Material found in the bog, along with the large number of burial sites, suggests the Windover people lived in villages. Generally lying on their left sides, the skeletons show minimal signs of having experienced an unusual or harsh death. Woven fabrics, the oldest ever discovered in the western hemisphere, possibly the world, were draped over the bodies and pinned to the soggy earth with stakes, forming a type of cloth coffin. Made of sabal palm and saw palmetto, these complex mats suggest a population with enough free time and social organization to create weaving bees.

The Windover Bog site has been allowed to revert to its natural state and remains inaccessible to the public. The overall story of its excavation, preservation, and analysis, however, is on permanent display in the Brevard Museum of History and Natural Science (2201 Michigan Avenue) in Cocoa. Full-scale replicas give evidence that even though Henry Flagler is properly identified as an industrial pioneer of the contemporary Sunshine State, the Windover people were the true snowbird trailblazers of the east coast of Florida.

Intricately incised bird bones found exclusively in the graves of women encourage scientists to investigate certain aspects of the Windover community, such as gender roles and the division of labor. Bone is 8 centimeters long. —Courtesy of Patty Meyers, director of the Brevard Museum of History and Natural Science

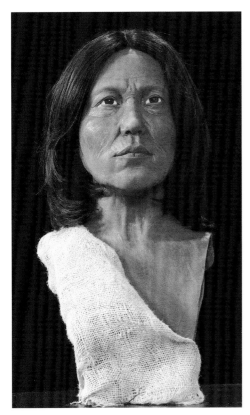

This sculpture of a typical 5-foot-2-inch-tall Windover woman is based on a forensic reconstruction of skeletal remains. —Courtesy of Patty Meyers, director of the Brevard Museum of History and Natural Science

Today, the Windover Bog site remains in almost the same condition as when discoveries were made in the early 1980s, one characterized by oxygen-free, low-acidity conditions ideal for preserving skeletal remains. —Courtesy of Bill Walls

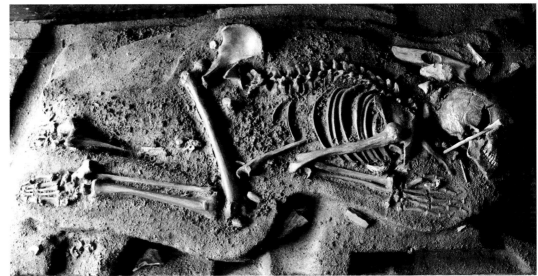

Although actual remains cannot be displayed, Windover replicas, such as this 7,000-year-old burial, are central exhibits in the Brevard Museum. —Courtesy of Patty Meyers, director of the Brevard Museum of History and Natural Science

22. Tibbs Bridge, Georgia

34° 44′ 11″ North, 84° 51′ 26″ West

Two unassuming trilobites that provide linkages to a world they once dominated.

Of the countless examples of ancient life that make up the world of paleontology, trilobites may well be archetypal, the model after which all other fossils species were patterned. Recognized as the biologic definition of the Cambrian Period—the Age of Trilobites—they are the undisputed first alpha invertebrate critters. The birth and death dates of class Trilobita bookend an era of history that begins with the Cambrian explosion, when Earth experienced an unprecedented acceleration in evolution, and closes with the Permian extinction, when as much as 96 percent of all marine species catastrophically expired.

Trilobite remains have been discovered on every continent. More than twenty thousand different species are known, a number that distinguishes them as perhaps the most diverse class of extinct organisms. Most were less than 4 inches long, but a few achieved lengths of 3 feet. The more common midsized species scavenged for benthic life, such as worms, by filtering seafloor mud. Larger varieties sought swimming victims, and a few may have adapted the nasty habit of cannibalism. Like modern-day crabs and lobsters, trilobites periodically shed their exoskeleton as they progressed from one stage of life to another, a process known as *molting*. The great majority of trilobite fossils are actually molts—outgrown and discarded exoskeletons. As suggested by their name, *trilobites* are made up of three body parts, a cephalon (head), thorax (abdomen), and pygidium (tail), segmentation that allowed them to coil into a ball when threatened by predators or agitated seas.

Trilobites played a significant role in the marine world for 289 million years, as the primary vacuum cleaners of the seafloor, until they became extinct in the last days of the Paleozoic Era. Acknowledged as the most abundant and diverse animals of Cambrian time, they have great value as index fossils—fossils that are geographically widespread and restricted to a very limited thickness (and thus limited time frame) of sedimentary rock. Geologists use several trilobite species worldwide to subdivide Cambrian time into ten geologic time units, each identified by the appearance of one or more index fossils. Specimens related to this time-and-fossil relationship can be observed within the accessible, thinly bedded Conasauga Formation along the east side of the Conasauga River at Tibbs Bridge, on Tibbs Bridge Road, about 3 miles southwest of Spring Place.

The two genera of Cambrian trilobites found at Tibbs Bridge, *Glyptagnostus* and *Aphelaspis*, are especially useful as index fossils. The former is recognized throughout the world—from Siberia to Kazakhstan to Australia—as undeniable and indisputable evidence that the host rock is Paibian in age, a 3-million-year segment of Cambrian time that occurred 497 to 494 million years ago. In contrast, *Aphelaspis* is useful for identifying Paibian-age rocks only in North America.

Many fossil collections begin with the unexpected sighting of a trilobite specimen in the rubble of a roadside exposure of limestone, siltstone, or shale. That said, at first glance, the index fossils of Tibbs Bridge are unimpressive: small, iron oxide stained, and delicate to the touch. With further consideration, however, their value as evidence for geologic context becomes noteworthy. They prove that the rock before you, the Conasauga Formation, is at least 494 million years old. On top of that, because they are the very oldest of catalogued trilobites, they could well change the merit of any collection, from one of mediocrity to one blessed with blue-ribbon status.

Bean-sized impressions of Aphelaspis brachyphasis, *accentuated by iron oxide staining, are common within the mudstones at Tibbs Bridge. This locale offers the southeasternmost exposures of this late Cambrian index fossil in the Appalachian Mountain region. Dime for scale.*

Broken and greatly disturbed, yet recognizable as fossil imprints, accumulations of Aphelaspis *at Tibbs Bridge are believed to be evidence of mass mortality events caused by mudflows cascading down the continental shelf of the late Cambrian ocean that inundated northwestern Georgia. Dime for scale.*

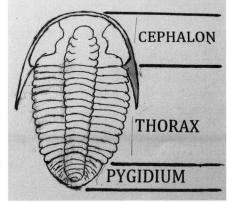

Fossil impression in Conasauga Formation limestone shows the broken stub of a cephalon spine (to the immediate left of dime, and colored red in the Aphelaspis *diagram). The spine is an identifying feature of* Aphelaspis brachyphasis. *Dime for scale.*

23. Makauwahi Cave, Hawaii

21° 53' 18" North, 159° 25' 08" West

A petrified system of sand dunes harboring a record of colonization, fossilization, and extinction.

About 5 million years ago, the plume of molten rock that had underlain the central section of the Pacific Ocean for 66 million years awoke from a period of quiescence, belched violently, and vomited basaltic magma onto the ocean floor. Over several hundreds of thousands of years the deep-seated turmoil continued, until finally the ever-growing mass of submerged volcanic material broke the ocean's surface to form Kauai, the Garden Isle of Hawaii.

Since then, eons of rain and wind have eroded the 1,000-cubic-mile mass of Kauai, resulting in emerald-carpeted valleys and Grand Canyon–like gorges, many capped with towering waterfalls. Pounding surf and tsunami-fueled waves have sculpted soaring cliffs, scalloped bays, and tapering points of land along its shoreline. The vast majority of the island is composed of dark fine-grained lava; however, here and there are small enclaves of sedimentary rock. The most unusual of these pockets lies on the southeast coast, in the Maha'ulepu Valley.

Lauded as the largest limestone cavern in the Hawaiian archipelago, Makauwahi Cave traces its origin to a field of cross-bedded dunes formed as wind deposited sand-sized particles of calcium carbonate eroded from Pleistocene-age coral and bryozoan reefs. After rainwater percolated through the dune sand and chemically compacted it, groundwater slowly dissolved portions of the resulting limestone, forming a 300-foot-long maze of interconnected caverns. Finally, some 7,000 years ago, a large section of the cavern roof collapsed, forming a sinkhole that is widely acknowledged as the richest fossil site in the Hawaiian Islands.

Digging into the lake, swamp, and storm-tossed sediment that fills the sinkhole is like reading a history book backward, starting on the last page and ending on the first. Beneath a clay-bound accumulation of contemporary litter lies a layer of sandy sediment containing a collection of bone fishhooks, iron nails from tall ships, and goat teeth. These items chronicle the arrival of Captain James Cook in 1778, en route to the hoped-for discovery of the Northwest Passage. An underlying jumble of boulders, sand, and gravel is interpreted to be evidence of a catastrophic tsunami wave that inundated the islands 400 to 500 years ago. Finally, there is a mother lode of fossils in a 30-to-35-foot-thick, coal-black layer of muck and peat. This layer is a treasure trove of fish and bird bones, leaves, pollen, spores, shells, and seeds, all perfectly preserved in the neutral pH environment formed by the chemical interaction of alkaline limestone and acidic groundwater.

Forty-five species of birds have been recovered from Makauwahi Cave, nearly half of which are extinct. Several can only be classified as bizarre: the turtle-jawed moa-nalo, a flightless goose-like duck; the 5,305-year-old nocturnal, near-blind, and flightless Kauai mole duck; and Pila's palila, a species of Hawaiian finch believed to have sported red, gold, and white feathers.

Makauwahi Cave, open to the public on a scheduled basis, houses evidence of three stages of historical extinction. The first was initiated 1,000 years ago with the arrival of Polynesians, who overhunted flightless birds. Further losses are attributed to the agricultural economy of a growing native population and the resulting decline of animal habitat. Finally, the arrival of Europeans and their invasive alien fauna spelled out the final chapter of a once-flourishing and diverse assemblage of midocean, tropical island life.

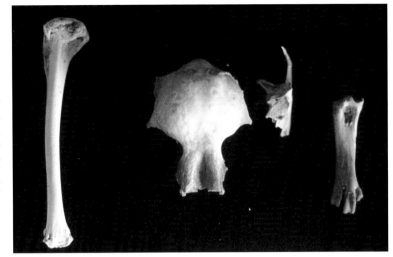

Bones of Talpanas lippa, *the Kauai mole duck, including the only skull ever found. This extinct, near-blind, and flightless bird, endemic to the island of Kauai, probably used its senses of touch and smell to search for food. Skull is 3 centimeters wide.*
—Courtesy of David Burney

Looking toward the entrance to Makauwahi Cave from a vantage point on the rim of the massive sinkhole. Wind-deposited carbonate sand eroded off nearby coral and bryozoan reefs to form the extensive cross-bedding to the right of the entrance. —Courtesy of David Burney

24. Clarkia Fossil Bowl, Idaho

46° 59′ 29″ North, 116° 16′ 35″ West

Fifteen-million-year-old fossils reminiscent of present-day autumn leaves.

Individuals attuned to the varied ramifications of nature easily recognize the evidence associated with the onset of each new season: the springtime budding of trees, the summertime blooming of milkweed, the color change of leaves in fall, and the first snowflakes. Unfortunately, these telltale manifestations of change seldom become part of the paleontological record because, once created, their demise is quickly assured by aging to mature green tones, consumption by monarch butterfly larvae, decay, and alteration to a liquid state, respectively. An exception to this rule exists in a bluff of interlayered silt, clay, and volcanic ash found off Idaho 3, 1.6 miles south of Clarkia. Here, autumnal effects from 15 million years ago can be unleashed by splitting open layers of laminated rock with handheld tools.

During the Miocene Epoch, flood basalts erupting from volcanoes associated with the building of the Columbia Plateau dammed the predecessor of the Saint Maries River, forming Lake Clarkia, a long and narrow reservoir that ranged in depth from 300 to 900 feet. Back then, a climate resembling that of the Everglades warmed north-central Idaho, and a canopy of flora similar to that growing today in the forests of central China and along the slopes of the southern Appalachian Mountains covered the region.

Nurtured by wet, hot summers and mild, dry winters, dense growths of deciduous and evergreen trees populated the forests of Lake Clarkia. Within the catalogue of needle and leaf fossils are those of the gymnosperm family, including incense cedar, Chinese flower pine, spruce, Chinese water pine, dawn redwood, and red cedar, and numerous angiosperm types, such as tulip poplar, alder, chestnut, sweet gum, sycamore, oak, beech, and holly. The material of more than 140 species have been extracted from Lake Clarkia rocks.

Considered one of the world's richest fossil finds, the Clarkia Fossil Bowl is classed as a *lagerstätte*, an exalted status given to sites with an unusual diversity of abundant, beautifully preserved fossils. For example, some sections of the Clarkia strata contain, on average, seven separate leaf specimens in a 6-by-6-inch area. Deposited in deep, cold, oxygen-deficient waters, the fossils are the remains of actual leaves, not the physical or organic impressions often found at other floral fossil sites. Paleontologists have developed techniques to break the bond between leaf and rock matrix using a water bath, to lift the separated three-dimensional artifact, and to preserve it within an acetate envelope.

When first exposed, Lake Clarkia leaves often display their longtime-ago autumn colors of yellow, red, and brown, and occasionally the summer tones of green. As the specimens are exposed to the degrading effects of twenty-first-century oxygen, however, the colors fade within a matter of seconds. The exceptional state of preservation of some specimens has allowed scientists to isolate small amounts of DNA, the chemical substance that contains the genetic instructions all living organisms employ in their development.

Regardless of the seasonal conditions that exist on any particular day or month of the year at the fee-based, family-operated Clarkia Fossil Bowl, one can continually find the earth tones of autumn in the three-dimensional, multihued leaves locked away within protective layers of rock, awaiting exposure after 15 million years of confinement. A word of caution, however: Stay alert and catch the moment—before the colors of antiquity are lost forever.

Maple trees originated in central China 110 million years ago and reached their widest distribution in the Miocene Epoch, the time when this winged seedpod from the Clarkia area became a fossil before it could pass along its DNA to future generations. Quarter for scale.

Dawn redwood (left) and bald cypress (right) from northern Idaho Miocene age rock.
—Courtesy of Bill Rember

Although 15 million years old, this birch leaf is so perfect it could have fallen from a tree yesterday.
—Courtesy of Bill Rember

Clarkia specimens of sweet gum leaves and pods (upper right) show that this tree has not changed one iota in 15 million years. —Courtesy of Bill Rember

This plethora of Clarkia Fossil Bowl plant fossils last felt the warm rays of the sun when rhinoceroses, camels, and prehistoric horses roamed North America. —Courtesy of Rachael Garrity

25. Hagerman Fossil Beds National Monument, Idaho

42° 49′ 04″ North, 114° 53′ 54″ West

Perhaps the most significant horse fossil site on Earth.

The wild horse population of North America covers a surprising collection of ranges and habitats. A herd of nearly 600 pony-sized animals roam Sable Island, a 13-square-mile expanse of sand and wind that lies 100 miles offshore of Nova Scotia. To the south, some 400 bankers, a breed possessing a friendly and docile personality, inhabit the Outer Banks of North Carolina. Called mustangs in the western United States, feral horses numbered more than 2 million a century ago, but today they number less than 50,000. The ancestors of these icons of the American frontier were brought to the New World by conquistadors, beginning with Hernán Cortés in 1519. Prior to that the horse was unknown to the many Native American tribes inhabiting North America.

The fossil history of horses dates back some 55 million years, long before any human existed. Their thundering hooves were silenced in North America about 10,000 years ago, victims of climate change and perhaps overhunting by prehistoric humans. A new chapter in the history of the horse, however, opened in 1928 when a rancher discovered horselike bones in 600-foot-high bluffs along the Snake River near Hagerman. Within a year Smithsonian Institute paleontologists were probing the 160 individual layers of sand, silt, and clay in search of more treasure.

Deposited 3.5 million years ago by flood-stage rivers flowing into ancient Lake Idaho, the sediments of the Hagerman Fossil Beds National Monument harbor at least 600 sites from which scientists have excavated and identified more than 180 plant and animal species. The most famous is the Hagerman horse, the earliest known hoofed mammal of the *Equus* genus, which includes modern-day horses, donkeys, and zebras.

With a well-studied inventory of 20 complete skeletons and more than 120 skulls, the characteristics of *Equus simplicidens*, the Hagerman horse, have become well known. It was about the size of a modern-day zebra, standing some 4 feet at the shoulder on legs distinguished by a single toe. It weighed between 400 and 900 pounds. The Hagerman horse lived in wetlands studded by grassy plains and forest stands, nurtured by 20 inches of yearly precipitation, and enveloped in a climate 7 degrees warmer than today's. The horse lived in harmony with a variety of other Pliocene Epoch mammals, including peccaries, camels, ground sloths, saber-toothed cats, and hyena-like dogs. The horse became extinct in North America 10,000 years ago.

Why so many horse fossils have been found in one location has long been a subject of controversy. One hypothesis posited that the site was a watering hole conducive to old and injured animals, who then died. Now it is believed that an entire herd died as a result of a catastrophic event, possibly drought causing death through starvation and thirst.

Lauded as the world's richest collection of late Pliocene fossils, the Hagerman fossil beds can be viewed from the Snake River overlook on Bell Rapids Road—the route of the famous Oregon Trail—2.5 miles west of the intersection with US 30, 4.5 miles south of Hagerman. The historic fossil sites are open only to authorized personnel, but a varied collection of Hagerman fossils is well displayed at the visitor center, 221 North State Street in Hagerman. In 1988 the Hagerman horse was designated Idaho's official state fossil.

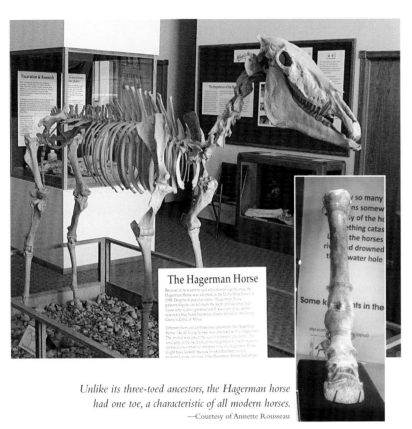

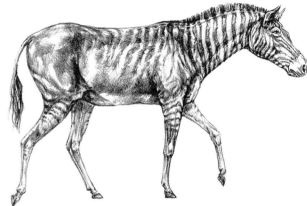

Regardless of its popular name, the Hagerman horse— Equus simplicidens *—is not a true horse, but rather a close relative to an endangered species of zebra living in the semiarid grasslands of Ethiopia and Kenya.*
—Courtesy of Annette Rousseau

Unlike its three-toed ancestors, the Hagerman horse had one toe, a characteristic of all modern horses.
—Courtesy of Annette Rousseau

Artist's interpretation of the Hagerman horse, believed to have been embellished with zebralike banding. —Courtesy of Annette Rousseau

You can view exposures of the Hagerman fossil beds (center) from the Snake River overlook, where the ancestral Snake River emptied into a large lake that was a long-term watering hole for a variety of animals between 8 and 2 million years ago.

26. Grafton Quarries, Illinois

38° 58′ 05″ North, 90° 25′ 17″ West

A historic trilobite graveyard.

The historical founding of the city of Grafton was most certainly not a spur-of-the-moment decision, and geology had a lot to do with it. The confluence of the Illinois and Mississippi Rivers offered convenient transportation options at a time when roads were practically nonexistent, and the floodplain was wide enough to accommodate the homes, buildings, and dock facilities required of a thriving river port. There were also massive cliffs of well-bedded, easily quarried dolomite: by the late nineteenth century, Grafton quarries were the principal supplier of stone, both facing and load-bearing rock, for almost all new construction in the flourishing downriver metropolis of St. Louis.

In the 1860s the quarries became the center of another, albeit curious, industry: quarrymen started selling trilobite specimens, nicknamed "rock dogs," to amateur and professional paleontologists. Over time more than twenty-five species were exchanged, but *Calymene celebra* was favored, a middle Silurian species relatively common in number and usually discovered in a complete form with uncurled posture.

For more than a century and a half the Grafton area has been one of the most celebrated Midwest collecting sites for trilobites. It's estimated that thousands of specimens have been removed from these rocks, the variety of which serves as prima facie evidence that everyday trilobites—both male and female—lived a regular and individualistic lifestyle: responding to periodic hormonal changes, they sought a protected site and, without further ado, molted.

At birth, supposedly after being hatched from an egg, a trilobite entered the protaspid (juvenile) period, the shortest of three stages of maturation it would experience in its lifetime. Both miniscule and spherical, it was clothed by a divided mineralized shield that would eventually develop a characteristic three-lobed morphology: a cephalon (head), thorax (abdomen), and pygidium (tail). The bifurcation into head and trunk anatomies preceded the meraspid (adolescence) stage, distinguished by the development of jointed segments that gave definition to a central thorax. Additional segments were added as the animal regularly molted, a process in which the creature shed its exoskeleton and grew a new one. The holaspid stage began once the trilobite had developed the number of thorax segments unique to its particular species. Now functioning as an adult, growth did not necessarily stop, as the fossil record shows that some animals eventually reached a state best described as trilobite obesity.

The process of molting, triggered when the exoskeleton became too formfitting for the animal to further develop its body, left the trilobite naked and quite vulnerable to predators, disease, or changes in environment. Nonetheless, some individuals produced as many as thirty exoskeletons, a reminder that many, if not most, trilobite fossils are merely discarded shells.

Trilobites are still found in the Grafton area. The community visitor center, on the eastern edge of town, is anchored in rock of an abandoned quarry that has yielded numerous examples. Exquisite fossils have been found in the rock rubble lining the hollows that drain the dolomite bluffs overlooking downtown Grafton. Pere Marquette State Park, 6 miles west of Grafton along Illinois 100, offers an informative display of area fossils. Although picked over by generations of fossil enthusiasts, the Grafton area retains its reputation as a treasure trove of museum-quality trilobites. The prized granddaddy of them all, however, is, and will continue to be, the celebrated *Calymene celebra*—king of the rock dogs.

├─── 1 inch ───┤

Adopted as the state fossil of Wisconsin in 1986, Calymene celebra *crawled on the ocean floor in a sluggish manner, searching for food with eyes that were somewhat smaller than those of the average Silurian-age trilobite.* —Courtesy of Kendall Hauer, Miami University

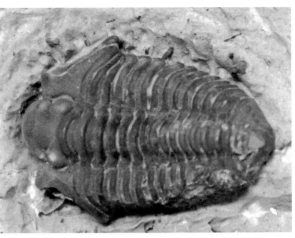

├─── 1 inch ───┤

Often found in a convex-up position, an indication of death in a low-energy environment that allowed the specimen to remain intact, Calymene niagarensis *is one of more than twenty-five species of trilobite found in the Silurian rocks of Illinois.* —Courtesy of Kendall Hauer, Miami University

├─── 1 inch ───┤

Many Grafton-area trilobites are found intact, but on occasion broken material—commonly known as fossil hash—is found. This small slab of limestone contains the disarticulated remains of Encrinurus, *another Grafton-area trilobite.* —Courtesy of Kendall Hauer, Miami University

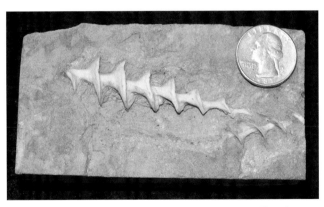

Grafton strata also host a variety of marine life of the middle to late Paleozoic Era, such as the bryozoan Archimedes. *Quarter for scale.*
—Specimen from the collection of Radford University Museum of the Earth Sciences

27. Mazonia-Braidwood State Fish and Wildlife Area, Illinois

41° 12′ 56″ North, 88° 16′ 09″ West

Concretions, nodules, and blobs add up to an amazing American lagerstätte.

The German word *lagerstätte* translates as "place of storage." Employed mainly by paleontologists, it is an eleven-letter message describing a supermarket locale of fossils that is exceptional in terms of quantity, variety, and degree of preservation. Lagerstätten developed where environmental conditions simultaneously inhibited decomposition and maintained the near-natural state of creatures upon death. Some sixty lagerstätten are known worldwide, including the famous Burgess Shale of Canada and the Solnhofen Limestone of Germany. Of no less significance is the Mazon Creek lagerstätte, centered on the junction of Kankakee, Will, and Grundy Counties and lauded as one of the most remarkable soft-bodied fossil deposits on Earth.

Some 307 million years ago, rivers draining the western slopes of the Appalachian Mountains emptied into a shallow-water sea whose shoreline lapped across the northeastern part of Illinois, then located several degrees south of the equator. During a decades-long episode, catastrophic monsoonal flooding, a rise in sea level, and the deposition of mud on the order of 3 feet per year—all associated with climate change and possible toxic algal blooms—caused a pandemic and the quick burial of the resident flora and fauna. Bacterial decomposition produced carbon dioxide gases that combined with groundwater-borne iron to form a nodule of siderite around each dead specimen. These coffin-like concretions prevented further decay and preserved the creatures' delicate structures. Button to dinner plate in size, the concretions are the Cracker Jack prize that has drawn paleontologists to the Mazon Creek lagerstätte for well over one hundred years.

Mazon Creek concretions are principally found within mounds of waste rock that date to the era of surface coal mining, which began in the 1920s. One massive pile is located west of Essex, the other, an amalgamation of eight piles, lies north of Braidwood. The Braidwood biota is terrestrial in nature: spiders, scorpions, horseshoe crabs, insects, and plant remains consisting of club mosses, horsetails, ginkgoes, and ferns. The Essex biota, in contrast, includes animals that lived in a marine to brackish nearshore environment: fish, worms, shrimp, snails, clams, and numerous specimens of mushroom-shaped jellyfish. The latter are lovingly called "blobs" because they lack diagnostic features.

The fact that some 40 percent of concretions are barren of fossils and thus considered duds is not an impediment to the anticipation of discovering the elusive and extinct *Tullimonstrum gregarium*, the most famous of all Mazon Creek fossils. Discovered in 1958 and designated Illinois's state fossil in 1989, "Tully" was a soft-bodied, 3- to 14-inch-long marine invertebrate that sported a long proboscis tipped with a jawlike appendage lined with eight small, sharp teeth. Not surprisingly, Tully is commonly classified as *incertae sedis*, scientific code for "Who knows what it is."

A no-cost permit, available at the Mazonia-Braidwood State Fish and Wildlife Area visitor center in Braceville (7705 East Huston Road), is license for the personal daily removal of a 5-gallon container of Essex-type concretions. For a nominal fee, any number of nodules can be culled from the spoil piles of the Fossil Rock Recreation Area in Wilmington (24615 West Strip Mine Road). You can open concretions by gently tapping along their edge or by placing them outdoors in water-filled buckets over the winter season, after which cycles of freezing and thawing will have hopefully caused them to split.

Two types of Pennsylvanian-age fern fossils have been found at Mazon Creek: true ferns and seed ferns. The former reproduced using tiny spores, whereas the latter, represented here by the genus Neuropteris, *used large seeds. Quarter for scale.*
—Courtesy of Radford University Museum of the Earth Sciences

Distantly related to modern-day spiders and scorpions and distinguished by an ancestry that dates back to the Cambrian Period, Euproops danae *is a Braidwood-biota horseshoe crab believed to have lived part of its life on land.* —Courtesy of Kendall Hauer, Miami University

Impressions of Pennsylvanian-age Essexella asherae. *This species constitutes 42 percent of all fossil finds in the Essex biota strata. Like the modern-day jellyfish, this species may have used stinging cells to capture small organisms. Dime for scale.*

28. Falls of the Ohio, Indiana

38° 16' 37" North, 85° 45' 49" West

This historic, cascading speed bump in the Ohio River is world famous for its fossils.

On October 20, 1811, Nicholas Roosevelt, his pregnant wife, and a crew of nine left Pittsburgh bound for New Orleans aboard his 120-foot-long side-wheeler, the first steamboat to attempt the 2,000-mile trip. They made it in 83 days, annoyed by a 6-week delay waiting for the river to rise enough for passage over the Falls of the Ohio, a 26-foot-high, 2.5-mile-long maelstrom of whitewater located 120 miles downstream from Cincinnati. Maintained by a protruding exposure of Jeffersonville Limestone, the falls were noted as early as the 1600s by French fur traders, both as an obstacle to navigation and as the site of beautifully exposed fossil beds. Today, a system of dams and diversionary canals tame the cascades, and the Falls of the Ohio State Park, open daily, protects the fossil beds.

Advertised as the most diverse fossil assemblage in North America, the falls inventory comprises more than 600 species of Devonian life. It lies scattered across a 220-acre rock exposure, part of a 1,000-mile-long, 387-million-year-old patch reef that can be traced as far north as Ontario. A full 41 percent of this inventory is composed of corals, bottom-dwelling marine animals that evolved in the Ordovician Period and still exist today. The most diverse group of corals found at the falls are classed as Rugosa, an extinct order that dominated the Devonian seas.

Many paleontologists are enamored by the morphology of rugose corals, so named because of their wrinkled, old-age appearance. Known also as horn corals due to a resemblance to the horn of a goat, rugose corals became extinct in Permian time. Though more than forty rugose genera are found at the falls, three—*Eridophyllum*, *Heliophyllum*, and *Siphonophrentis*—are of particular interest, as they were instrumental in addressing a question that had pestered investigators for decades: Do rugose corals have any relation to small and large-scale changes in astronomical phenomena?

It seemed the key to an answer must be related to the wrinkles constituting rugose complexion. Paleontologists speculated they were growth ridges, increments of calcium carbonate production spurred by cyclical fluctuations in temperature. They expected there would be 365 (the number of days in an Earth year) bookended by supposed annual rings. Attempts to count the wrinkles were laden with difficulty. Storms, predator attack, and environmental change all affected their development, and even after death, ridge detail could be altered by abrasion and corrosion.

Success was eventually achieved with the assembly of museum-quality specimens of rugose corals representing the same genera found in quantity at the falls. Wrinkle counts were very carefully made and the results published: Devonian rugose corals customized their exoskeleton by adding, on average, some 400 wrinkles per year, not the expected 365. This unanticipated evidence indicated that the number of days in the Earth year had decreased with the passage of geologic time. Subsequent studies confirmed this decrease: Cambrian-age corals gave an average count of 418 days per year, and those from the Late Pennsylvanian Period 385.

Calendar specifics for the 400-day-long Devonian year are now generally acknowledged: it lasted for 13 months, each month approximately 31 days long, and each day less than 22 hours. Earth's rotational velocity is decelerating at the rate of 2 seconds every 100,000 years, a slow braking action attributed to the frictional force caused by twice-daily tides. The future can now be stated in terms of yet another "wrinkle"—Earth will eventually come to a complete stop, one side eternally cold the other side eternally hot.

Heliophyllum *is a common solitary rugose coral found at the Falls of the Ohio. Penny for scale.* —Courtesy of Kathi Mirto

Rather than living a solitary life, Eridophyllum *corals lived in groups or mounds. Foot-long ruler for scale.* —Courtesy of Indiana Department of Natural Resources, Falls of the Ohio State Park

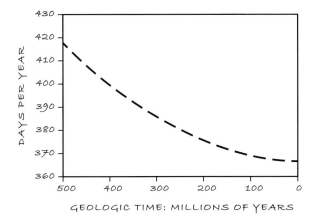

Careful counting of the growth ridges of museum-quality rugose corals indicates the number of days in Earth's year has decreased approximately 53 days over the past 500 million years as a result of tidal friction, from 418 during Cambrian time to 365 today.

The V-shaped growth of wrinkled Siphonophrentis *rugose corals (black), commonly called "petrified buffalo horns" by early settlers due to their shape, intertwined with nonrugose corals (center). The black coral at right center is 16 inches long.* —Courtesy of Indiana Department of Natural Resources, Falls of the Ohio State Park

29. Whitewater River Gorge, Indiana

39° 48′ 26″ North, 84° 54′ 33″ West

A 3.5-mile-long stroll across the fossil seafloor of the Tippecanoe Sea.

Many college freshman geology textbooks contain the illuminating concept that a marine invasion of gargantuan extent flooded the American landscape during Late Ordovician time. Statements that run the gamut of literary expression support the scientific understanding that this oceanic transgression was indeed the most invasive continental flood experienced by any landmass in geologic time. The only dry land interrupting the horizon-to-horizon aquatic vista was Taconica, a chain of volcanic islands that had formed along the east coast of proto–North America during the Taconic orogeny, the first of three mountain-building phases that developed the Appalachian Mountains during the Paleozoic Era.

Organisms thrived in the incubating environment of the warm, shallow, oxygen-rich water of this Tippecanoe Sea. Mud was the deposit of choice offshore, while soupy and slimy ooze, sediment composed of at least 30 percent calcareous skeletal remains, accumulated in clearer, inshore waters. Algae helped produce an abundance of carbonate sediment that, in turn, acted as the burial ground for an extensive suite of invertebrate fossils, considered by many paleontologists to be evidence of the most dramatic increase in biologic diversity recorded in geologic history. More than seven hundred species of life have been recorded in the Ordovician-age strata exposed in the tristate Indiana, Ohio, and Kentucky region. Fossil aficionados have suggested that if all the fossils were removed from these rocks, the area would surely slump below sea level.

For decades geologists, paleontologists, and paleobiologists have traveled to the city of Richmond to collect a sampling from this plethora of well-preserved trilobites, bryozoans, brachiopods, pelecypods, echinoderms, and graptolites of Late Ordovician age. The widespread fame of these fossils is evidenced by the fact that geologists use the term Richmondian to refer to sedimentary rock of the same age throughout all of North America.

Exposures of Richmond-age strata are conveniently exposed along the 3.5-mile length of the Whitewater Gorge Trail, which extends through the heart of the city from Test Road north to Thistlethwaite Falls. The physical condition of the fossils depends on their environment of deposition. Delicate and unbroken specimens are associated with the Waynesville Formation, shale strata deposited offshore, whereas battered and broken fossils clearly identify limestone beds of the Whitewater Formation, deposited in shallow nearshore water that was agitated by waves and storms. A mixture of broken and whole fossils is suggestive of the Liberty Formation, deposited in a transitional environment. Geologists can identify these formations solely by the index fossils they contain. Waynesville strata exclusively harbor the brachiopods *Strophomena nutans* and *Tetraphalerella neglecta*, whereas the Whitewater Formation contains the distinctive brachiopod *Platystrophia acutilirata*.

As the last of the ice age glaciers melted northward from the Richmond area 20,000 to 18,000 years ago, one unusually large lobe of ice lingered, temporarily protected by its bulk and isolation. Eventually it succumbed to the warming climate, producing floods of sediment-laden meltwater that scoured a deep, narrow channel—the Whitewater Gorge—across the Ordovician bedrock, exposing fossils to be discovered by weekend paleontologists.

Richmondian fossils can be found just about anywhere rock is found in Whitewater Gorge. Two sites along the Whitewater Gorge Trail are especially accessible: the nearly vertical cliffs north and south of where Bridge Avenue crosses the river, and at the base of Thistlethwaite Falls, the remnant of a waterfall that has been in the process of migrating upstream for thousands of years.

1 inch

The common horn coral Grewingkia canadensis *can be found in all three Richmond-age formations: Waynesville, Liberty, and Whitewater.* —Courtesy of Kendall Hauer, Miami University

1 inch

Arguably the most sought-after and treasured of all Ordovician-age fossils, specimens of the trilobite Isotelus maximus *are found in all three Richmond-age formations.* —Courtesy of Kendall Hauer, Miami University

The presence of the index fossil Platystrophia acutilirata, *a brachiopod, is evidence the host rock is part of the Whitewater Formation.* —Courtesy of Kendall Hauer, Miami University

1 inch

A specimen of the brachiopod Tetraphalerella neglecta *is proof positive the host rock is the Waynesville Formation.* —Courtesy of Kendall Hauer, Miami University

1 inch

30. Devonian Fossil Gorge, Iowa

41° 43′ 18″ North, 91° 31′ 55″ West

Annals of deep time catastrophically exhumed by contemporary forces of nature.

Many hydrologists consider the Great Flood of 1993 to be the most devastating to have occurred in the United States. The statistics are staggering: 50,000 homes destroyed or damaged; 30,000 square miles of land flooded, many for nearly 200 days; 60,000 people displaced and an estimated 50 lives lost; 1,000 levees topped or failed; and damages that exceeded $15 billion.

In east-central Iowa, 48 inches of rain fell during the 5-month period beginning on April Fool's Day, a deluge compared to the average 33 inches that falls in a normal year. On top of this, it fell on soil with above-normal moisture levels and on abnormally full reservoirs. Heavy winter snowfall and persistent spring storms forced the weather service to issue more than one thousand flood alerts for the central plains. Reports of the Mississippi River reaching levels 20 feet above flood stage—the highest recorded in 228 years—only added to fears of a pending disaster, a situation later declared a once-in-a-300-year occurrence.

Conditions were not any better at Coralville Lake, 3.5 miles north of I-80 exit 244 in Iowa City. By July 5, the lake's water had risen 4.5 feet above the crest of the spillway and was surging downstream at a rate of 130,000 gallons per second. Within hours all vegetation, topsoil, pavement, sediment, and loosely consolidated bedrock had been stripped away from a 2-acre rectangle of land. The overflow and catastrophic erosion continued for twenty-eight days, but it had one positive effect: it created the Devonian Fossil Gorge, a park maintained by the US Army Corps of Engineers.

Bedrock at the gorge is composed of limestone of the Little Cedar Formation, deposited during Middle Devonian time, when ancestral North America straddled tropical latitudes south of the equator. Buried for 385 million years and exhumed by the flood, the strata allow visitors to return to the distant past, when the Hawkeye State was inundated by a balmy, shallow-water sea that stretched from horizon to horizon. A walk across exposed limestone is, in effect, a stroll through centuries of Devonian time.

Many types of fossils are present in the limestone: brachiopods, crinoids, bryozoans, cephalopods, and elusive trilobites. Two fossils, however, are remarkable: a spectacular coral and the remains of an extinct predator that ruled the Age of Fishes. *Hexagonaria* is an extinct colonial coral associated with Devonian-age rocks worldwide. Rock hounds and jewelry makers have long treasured this striking coral, with its intricate array of adjoining six-sided chambers, each of which housed a singular animal. When polished to a sheen, they become objets d'art, ranging from paperweights to pendants. The *Hexagonaria* specimens in Devonian Fossil Gorge are part of a massive reef that extends east to the Mississippi River and north to the Minnesota border.

As for the predator that ruled the Devonian seas, *Dunkleosteus* was probably a relatively slow and laggardly swimmer because of the heavy armor that protected its front end. What this 35-foot-long, 4-ton alpha marauder lacked in finesse, however, it possessed in might: its jaws could exert 80,000 pounds of force per square inch. Instead of teeth, this gluttonous hunter used sharp bony plates that sliced with equal effect through both flesh and bone. These primitive, predatory fish terrorized the Paleozoic seas for more than 70 million years and would, on occasion, cannibalize each other.

Hallmarked by twenty "discovery points," Devonian Fossil Gorge is an informative deep-time depository of Devonian life, a must-visit exemplar created by modern-day erosional forces of unprecedented magnitude.

The characteristic knobby surface of a dorsal plate of Dunkleosteus, *on view along with other fossil specimens at the Devonian Fossil Gorge visitor center. Pen for scale.*

Considered the most spectacular of Devonian Fossil Gorge brachiopods, Spirifer *specimens are easily recognized by their elongate shape (broken example in center and rock-embedded specimen at left). Other commonly found brachs are smaller in size and more rounded in form (lower right). Pen for scale.*

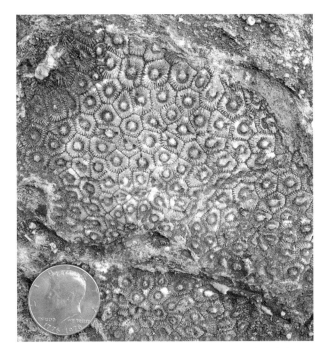

Close-up of Hexagonaria, *a colonial coral constructed of an ideal array of six-sided chambers, each of which housed an individual animal that used a cluster of tentacles to catch food and sting threatening organisms. Half-dollar for scale.*

Devonian Fossil Gorge, seen downstream from the Coralville Lake spillway, is composed of shelves of sedimentary rock that originated as fine-grained lime-rich deposits on the floor of an inland sea 385 million years ago.

31. Fossil and Prairie Park Preserve and Center, Iowa

43° 02′ 50″ North, 92° 58′ 51″ West

Collect a full quart-jar of brachiopods in an hour.

Populations across America have long taken pride in the belief that they live in one-of-a-kind locations: the birthplace of a noteworthy individual, the site of a historic battle, or the hometown of a championship athletic team. Just as the subjects of local pride may differ, so can the means of boasting. Some communities favor light-pole banners, some courthouse statues, and others bumper stickers. The crossroads village of Rockford has chosen rhomboid and oval, pole-mounted, all-weather signs to bear their simple message: "Rockford Home of Devonian Fossils." These five words, although somewhat fuzzy in meaning, spell out the story of a community recognizing and preserving its world-class fossil-collecting locale.

Toward the end of the Devonian Period, about 375 million years ago, ancestral North America straddled the equator, almost entirely inundated by the Kaskaskia Sea, a shallow-water extension of the Iapetus Ocean. Throughout north-central Iowa fine-grained sediments were being deposited in oxygenated, sunlit environments filled with rich and diverse populations of invertebrate organisms. It was a time of geologic tranquility in the Hawkeye State. To the east, the Appalachian Mountains had reached the halfway point of a several-hundred-million-year episode of mountain building. To the west, the horizon was marked by the tidal fluctuations of a seemingly boundless ocean.

Fast-forward to the year 1910. The Rockford Brick and Tile Company began mining the blue-gray shale beds of the Lime Creek Formation to produce common bricks and agricultural drainage tiles. For six decades the operation prospered, even in the 1940s when some fifteen German prisoners of war replaced employees called to military service, but by 1977 it was nothing but a dump site. However, the community realized its value and supported its transformation to Fossil and Prairie Park Preserve and Center, a multipurpose educational facility that opened to the public in 2001.

The uppermost near-surface exposures in the quarry pits are composed of the Cerro Gordo Member of the Lime Creek Formation. These strata contain an unusually rich collection of fossils that weather out as distinct, museum-quality specimens that can be collected with ease. Bryozoans, gastropods, cephalopods, and crinoids are present, but Prairie Park has earned a reputation as a coral site of distinction and a brachiopod locale of excellence. Since they were first reported in the 1800s, literally tens of thousands of Cerro Gordo brachiopods have been collected by generations of paleontologists. Because of their abundance and outstanding state of preservation, universities and museums around the world have them on display. At least forty species have been found at Prairie Park.

Pachyphyllum woodmani, a colonial coral composed of banded, cylindrical calcium carbonate cups adorned with elevated rims, is the coral to find at Prairie Park. The surface of a pristine specimen is reminiscent of that of the moon, a landscape of impact craters characterized by raised margins. Found in rock as far afield as New York and Arizona, this coral provides supporting evidence that ancestral North America was extensively flooded during Devonian time.

Located 1.5 miles southwest of Rockford on County Road B47, the Fossil and Prairie Park Preserve and Center is a renowned site for collecting Devonian fossils. The education center is open only on a seasonal basis, but the adjacent fossil-collecting area is accessible daily from sunrise to sunset. Leave hammer and chisel at home, but do bring sunscreen and protective kneepads for this open-air operation, which is all about keeping your hands, feet, and eyes on the ground and taking home all the fossils you find.

A 1-inch-wide specimen of Cyrtospirifer whitneyi, *an extinct brachiopod of a genus that dominated the seas during Late Devonian time. Note the distinctive ornamentation lines that converge on the raised and centrally located beak.*
—Courtesy of Steve Uchytil

This sign beckons both amateur and professional paleontologists to visit Rockford to fulfill their fossil-collecting dreams.

Close-up of the surface of a typical Pachyphyllum woodmani *colonial coral found at the fossil park. In life, a singular coral occupied each of the raised 0.2-inch-wide cups that are accentuated by radiating ridges.*
—Courtesy of Steve Uchytil

Fossils are always plentiful in the highly weathered Lime Creek Formation, but the best time to visit the fossil park is after an extended rain has flushed new specimens to the surface.

32. Round Mound, Kansas

37° 48' 29" North, 96° 03' 27" West

An off-the-beaten-path site harboring pygmy-sized brachiopods and horn corals.

Fossil-collecting sites are like snowflakes: no two are alike. Some locales are well known and visited by hordes of enthusiasts who overwork them, whereas a treasured few are fossil rich in terms of quantity and variety but accessible only to individuals with a collecting permit. Many fossil settings are off-limits due to private or government ownership, others host a limited variety of specimens, some are difficult to reach, and a few are downright dangerous to visit.

And then there are all the others: localities that are family friendly, are reasonably accessible, and harbor a catalogue of fossils that, while not worthy of adding a new wing to the local natural science museum, are worth a balmy afternoon romp in the field. Round Mound fits this description. This 43-foot-tall, city block–long hillside outcrop intercepts the flatlands of southeastern Kansas 2.6 miles south of the intersection of US 54 and County Highway 17A (AA50), on the southeast side of Neal.

Composed of interlayered sequences of limestone and shale that make up the Kanwaka Formation, Round Mound strata were deposited 300 million years ago, during the latter part of the Pennsylvanian Period. This rural site was then situated in a warm, equatorial tidal flat comprising a combination of sand flats, mudflats, and river channels. The fossilized remains of at least forty different organisms have been identified at Round Mound, the most common being brachiopods, horn corals, crinoids, and bivalves (oysters and clams). Weekend visitors can easily collect representative examples of each, but success requires hands-and-knees examination, because the typical specimen is less than 0.5 inch in size.

Brachiopods are marine animals that usually attach themselves to rocks or other objects on the seafloor with a flexible stalk. The shell that encloses the soft parts of the animal consists of two bilaterally symmetrical valves, meaning the left valve is a mirror image of the right. A particular favorite of many paleontologists, "brachs" are common, can be conveniently identified by their external decorations of concentric wrinkles and radial ribs, are generally found in well-preserved form, and have a wide geographic distribution. Two types of extinct genera are found at Round Mound: *Composita* and *Chonetes*.

At least four species of *Composita* were identified in the Kanwaka Formation in early studies—*elongata*, *ovata*, *subtilita*, and *trilobata*—but recent analyses suggest they represent an evolving series of organisms whose valves varied slightly in form and structure due to life in different tidal-flat environments. The jury is still out on this discussion, a situation that is common in the field of paleontology. Less controversy is associated with *Chonetes*, a very common brachiopod in Pennsylvanian-age strata in Kansas. Often more that 75 percent of the invertebrate specimens of any fossil assemblage in the state is of this genus. Semicircular in shape and generally 0.5 inch wide, *Chonetes* characteristically possesses a series of microscopic nibs along the flat edge of one valve, the remnants of spines the animal employed to anchor itself to the muddy seafloor.

Horn corals were simple, solitary animals that secreted exoskeletons of calcium carbonate. As they grew older they became ever larger in diameter and heavier in weight, until their mass caused them to topple. Continued growth produced a distinctive hornlike bend. Half-inch-sized specimens of *Lophophyllidium*, ornamented with well-developed longitudinal ridges, are found in the shale of the Kanwaka Formation, especially after a vigorous rain has reworked the weathered strata.

Crinoid columns and individual columnals (circular fossils) are common at Round Mound, but the real prizes are the numerous and minute plates that make up the calyx, or central body, of the crinoid (triangle-shaped fossils in center). Penny for scale.

Stub-like projections along the edge of the extinct half-moon-shaped shell of Chonetes are the remains of delicate spines the brachiopod used to anchor itself to the seafloor. Penny for scale.

Nose-to-the ground exploration is mandatory if you hope to discover the horn coral Lophophyllidium, characterized by surface striations, but even closer scrutiny is needed to find the relatively scarce pea-sized brachiopod Hustedia (right). Penny for scale.

Two types of sedimentary rock make up the Kanwaka Formation at Round Mound: massive beds of battleship-gray limestone (center and lower), and thin beds of shale that weather to a pale yellow (center). Whole fossils are usually found within the shale beds.

Composita, an index fossil for Devonian-to-Permian-age strata, is found as far afield as the United States, Indonesia, and Peru. Penny for scale.

33. Sternberg Museum of Natural History, Kansas

38° 53′ 20″ North, 99° 17′ 59″ West

A dedicated fossil-hunting family who wrote an important chapter of American paleontology.

Perspiring profusely, body wracked from the ill effects of alkali water, and eyes obscured by clouds of lime dust, Charles H. Sternberg continued his quest to discover any complete skeleton within the Niobrara Chalk exposures that would become Gove, Logan, and Wallace Counties in Kansas. He finally struck pay dirt in the late spring of 1876 when he unearthed an intact mosasaur, later christened *Clidastes tortor* by his mentor Edward D. Cope, a preeminent paleontologist of the nineteenth century. One hundred and twenty-three years later, Fort Hays State University opened the doors of its Sternberg Museum of Natural History, a state-of-the-art facility dedicated to the fossil-famous Sternberg family. The fossil-exploration escapades and finds of this family constitute one of the most interesting chapters in the paleontological history of America.

Buoyed by his mosasaur discovery, Charles expanded his explorations into the rich Cenozoic-age fossil beds of northwestern Kansas, discovering a variety of elephant, turtle, and rhinoceros bones. Self-taught in the nature of fossils and the techniques of successful fossil hunting, early on he knew he wanted to be a field paleontologist searching and discovering amongst the rocks, not a university or museum researcher bound to a desk. His career, therefore, was to be one of continuing good and bad news: bad due to extreme weather conditions, the loneliness of remote locations, and the constant risk of illness and injury; good because virgin world-class fossil beds could still be found just beyond the next horizon in the last decades of the nineteenth century.

Charles's income was derived from a variety of sources, including working under contract to professional paleontologists and selling specimens from his private collection. His interests took him as far afield as Oregon, Montana, New Mexico, Texas, California, Wyoming, the dinosaur-bone deposits of Alberta, and the windswept Pampas of Patagonia in Argentina. In 1928 he assembled a collection of thirty-five giant Cretaceous-age ammonites in Baja California, left the field for good, and died fifteen years later at the age of ninety-three. In many ways Charles Sternberg's passing marked the end of the great age of fossil explorations in America, when paleontology was a youthful and controversial field of study and the eminent natural history museums of today were still in developmental stages.

Each of Charles's three sons made their own mark in the paleontological world. George and Levi electrified the scientific community with their 1908 discovery of a 66-million-year-old mummy-like specimen of a duckbill dinosaur, complete with skin impressions and ossified fibrous tissue. Forty-four years later George unearthed the famous fish-within-a-fish fossil, a defining and lasting moment of gluttony: enclosed within the rib cage of a perfectly preserved 14-foot-long *Xiphactinus*, a predatory, bony Cretaceous-age fish, was a 6-foot-long, partially digested tarpon-like victim. And as curator of the National Museum of Canada, Charles M. became a renowned expert on Canadian dinosaurs.

The fossil collections of Charles H. Sternberg are distributed in museums around the world. Many of the specimens collected by George are on display in the Sternberg Museum of Natural History (3000 Sternberg Drive) in Hays. The more than three million deep-time items housed here are a lasting testimony to the lifelong endeavors of a family of intrepid fossil hunters. Their legacy of paleontological specimens, on display and in storage at the museum, rivals that of any college or university in the world.

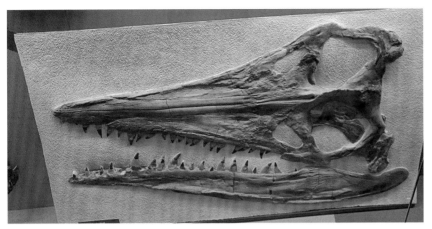

Unearthed in Kansas in 1950 by George Sternberg, this 5-foot-long Brachauchenius *skull is representative of the pliosaurs—large carnivorous marine reptiles—that prowled the great inland sea that inundated part of North America during Cretaceous time.* —Courtesy of the Sternberg Museum of Natural History

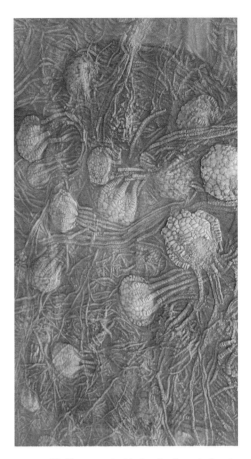

Platecarpus, *a mosasaur ancestor to the living Komodo dragon, abandoned its terrestrial life in Late Cretaceous time and evolved to life at sea by modifying its legs into flipper-like structures (bottom set of bones). Ranging in length from 3 to more than 40 feet, these marine reptiles became extinct 66 million years ago.* —Courtesy of the Sternberg Museum of Natural History

Unlike most crinoids that lived attached to the seafloor, Uintacrinus *floated freely and fed by extending its string-like "arms" to capture food. This colony died of some catastrophic event, perhaps smothered by volcanic ash or a turbidity current or poisoned by an algal bloom.* —Courtesy of the Sternberg Museum of Natural History

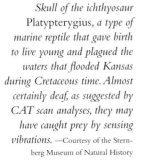

Skull of the ichthyosaur Platypterygius, *a type of marine reptile that gave birth to live young and plagued the waters that flooded Kansas during Cretaceous time. Almost certainly deaf, as suggested by CAT scan analyses, they may have caught prey by sensing vibrations.* —Courtesy of the Sternberg Museum of Natural History

34. Big Bone Lick State Historic Site, Kentucky

38° 53′ 20″ North, 84° 44′ 52″ West

The birthplace of American vertebrate paleontology.

A variety of enterprises have occupied the rolling lowlands of the Big Bone Creek valley, off Kentucky 338, 3 miles west of Beaverlick. An early effort involved the manufacture of salt by boiling the warm saline waters that bubbled to the surface. Around 1800 the construction of the fashionable Clay House, a spa designed to entice travelers to loiter, drink, and bathe in the sulfur-bearing springs, enhanced the area's allure. Today the area is home to the 813-acre Big Bone Lick State Historic Site, one of the most famous paleontological locales in North America.

Early accounts describe a network of 3-to-4-foot-deep paths the width of two wagons converging on the historic site's springs, which had been sculpted by generations of stomping and burrowing animals seeking a source of sodium to supplement their plant-heavy diets. Shawnee, Wyandot, Iroquois, and Delaware Indians also visited the area, hoping to kill a young, old, or diseased bison. The bison disappeared around 1800, and salt making ended in 1812, but long before then the site had accumulated a most unusual commodity: thousands of disarticulated bones described simply as big, bigger, and biggest.

In 1739 the Frenchman Baron Charles de Longueuil was the first to collect bones from the area, the highlights of which were a 40-inch-long femur, three molar teeth, and a tusk. He sent the collection to Europe for study. The analyses caused a flurry of interpretation: the femur and tusk were declared those of a gargantuan elephant, and one gigantic molar was incorrectly classified a hippopotamus tooth.

George Croghan, a Pittsburgh-based federal administrator for Indian affairs sent a second collection, labeled "elephant bones," to Europe in 1767. After examining them, Benjamin Franklin, then living in London, questioned their identification, noting modern African and Asian elephants lived in hot climates, whereas the American specimens were found in a cold-climate region. As the eighteenth century drew to a close, the fossils were a continuing source of dissension and controversy: Did the bones represent one or more species, extinct or extant, herbivorous or carnivorous, elephant or elephant-like? Many experts were bothered by mutterings of extinction, a concept alien to those who believed a benign God would never allow such a thing to happen to one of his creations.

Following his epic trip to the Pacific Ocean with Meriwether Lewis, William Clark visited Big Bone Creek in 1807, collected three hundred bones, and sent them to the White House for President Thomas Jefferson, an amateur paleontologist, to examine. Jefferson shared the collection with friends for further examination, one result being the introduction of the word *mastodon* into the controversy.

The Big Bone Creek valley is internationally recognized as a fossil boneyard of exceptional repute and the depository of the partial remains of at least eight extinct species of mammals. Some animals, such as the humongous mastodon, became entrapped in the spongy ground, and others died of old age or from having fought to the death over a favorite drinking site. The mastodon, Harlan's ground sloth, stag-moose, woodland musk ox, and ancient bison first became known to science on the basis of local discoveries. In addition, the remains of woolly and Columbian mammoths, complex-toothed horse, and Jefferson's ground sloth have also been recovered from the Pleistocene-age soils of the region.

After a rich history of scientific investigations exceeding 275 years, the Big Bone Creek valley continues to yield a treasure trove of large vertebrate bones, a dividend of erosion. These finds add substance to its reputation as the birthplace of American vertebrate paleontology.

More than big bones are found at the state historic site. At least twenty thumb-sized brachiopods occupy this 1-foot-long slab of Ordovician-age limestone. —Courtesy of Dean Henson, Big Bone Lick State Historic Site

Layers of elongated crinoid stems, two prized thecas (the animal's digestive and reproductive system), and food-gathering arms (lower left and center right) cover this museum-quality, 1-foot-wide slab of Big Bone Ordovician-age bedrock. —Courtesy of Dean Henson, Big Bone Lick State Historic Site

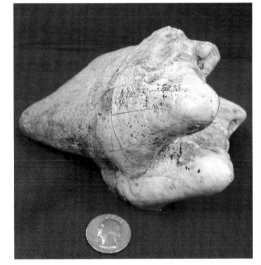

Early-nineteenth-century paleontologists, seeing a resemblance between the shape of mastodon molars and the human breast, coined the word mastodon, *meaning "breast tooth." Quarter for scale.* —Courtesy of Dean Henson, Big Bone Lick State Historic Site

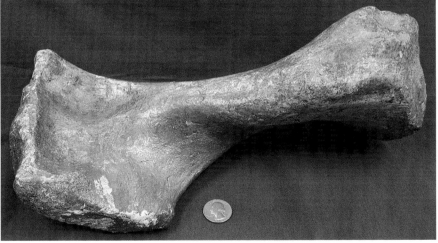

Adult mastodons grew 7 to 10 feet tall and weighed 4 to 6 tons. Individual skeletal parts, such as this 13-inch-long upper-arm bone of a juvenile, suggest the massive nature of these browsing and grazing forest-dwelling animals. Quarter for scale. —Courtesy of Dean Henson, Big Bone Lick State Historic Site

35. Danville Bryozoan Reef, Kentucky

37° 39′ 06″ North, 84° 52′ 29″ West

A rare bryozoan reef with crabgrass- and kudzu-style fossil features.

Beginning 3.5 billion years ago, simple, microscopic, soft-bodied masses of bacteria and algae flourished in primeval oceans for millions of years. The evidence for this life, however, is sparse, a void in the fossil record attributed either to the inability of the organisms to secrete skeletal material, or to the fact that following each organism's death its secreted hard parts dissolved in the aquatic environment. Then came change in the form of the Cambrian explosion, the most intense burst of evolutionary development ever known.

Commencing 541 million years ago, the sun rose on a world soon to become forever different. Over the course of the following 20 to 40 million years—seemingly overnight in the geologic sense of time—a wide variety of invertebrates erupted onto the scene. They were capable of secreting skeletal material that, following death, created a visual fossil record. Included in the Cambrian paleontological inventory are organisms that swam, crawled, burrowed, and scavenged: mollusks, brachiopods, early echinoderms, ancestors of spiders and insects, and the iconic trilobite. Representatives of almost every known multicellular category of organism capable of secreting skeletal material made an appearance, with one important exception—the phylum Bryozoa.

The Ordovician Period that followed was marked by the Tippecanoe Sea, a shallow water body that covered the interior of North America from Pennsylvania west to Nevada and north to the Hudson Bay. This inundation, considered the greatest-ever flooding of any landmass, combined with a tropical climate and elevated carbon dioxide levels, forged a nutrient-rich environment that encouraged diversification above and beyond that of the Cambrian. Within this favorable environment the bryozoan made its first appearance.

Commonly known as "moss animals," bryozoans are minute, dominantly marine animals that live in colonies attached directly to the seafloor, to organisms and objects on the seafloor, and to floating objects. Once established in environmental niches during the Ordovician, they evolved with results comparable in shape and design to kudzu and crabgrass, two types of plants that are the bane of modern-day gardeners and landscape architects. Some Ordovician bryozoans grew in leaf-, bush-, or fanlike forms resembling branched, twig-like colonies of crabgrass. Others slowly encrusted surfaces of rocks and shells kudzu-style, creating a matted, laminated, sheetlike surface. Both forms were composed of millions of microscopic *zooids*, sedentary, individual calcareous organisms that compose the colonies.

Under opportune conditions, bryozoan colonies grew in such abundance they became the principal builders of reefs, wave-resistant ridgelike structures. Central and northern Kentucky is famous for Ordovician-age reefs built in the warm 5-to-60-foot depths of the Tippecanoe Sea. A typical structure occurs at the intersection of US 150 and County Road 1822, about 5 miles west of downtown Danville. Inch-sized fragments of bryozoans cover limestone and shale of the Clays Ferry Formation in minimally disturbed, in situ positions that, with a little patience and as much glue, can be assembled into several-inches-high, branch- and moss-like reconstructions of the living colony.

Bryozoan communities have existed since Ordovician time, but reefs like this are not common. Studies indicate they normally grow at an average rate of 1 inch every 2,500 years. The 8-to-10-foot-high roadside exposure at Danville is, therefore, an opportune site to leisurely investigate both the crabgrass- and kudzu-style bryozoan colonies that flourished for millions and millions of years.

The kudzu-style of bryozoan formed a single layer of encrusting zooids on a variety of seafloor surfaces. This style of growth is found in relative abundance in the Danville reef. Green color added for emphasis. Dime for scale.

microscopic examination of a branching, crabgrass-style bryozoan reveals a series of les, each once the home of a zooid, a tiny filter-feeding animal with a mouth surunded by food-gathering tentacles. Red color added for emphasis. Penny for scale.

While a few brachiopods, cephalopods, and crinoids are found in the Danville reef, this site is famous for the presence of easily collected bryozoan fossils. Crabgrass-style specimens are reported to occur in a two-to-one ratio to kudzu-style samples. Penny for scale.

36. Grand Isle, Louisiana

29° 15' 40" North, 89° 57' 08" West

A myriad of sand-sized specimens makes up for the lack of macrofossils in Louisiana.

The location of fossil-rich roadside outcrops and field exposures is common knowledge among ancient-life enthusiasts in many American states, but the paleontology cache of Louisiana remains a topic seemingly shrouded in secrecy, and for good reason: fossil-bearing strata in the Pelican State are, in a word, rare. River basin, swamp, and wetland landscapes cover large portions of the state, creating environments unfavorable to the discovery of fossils. Making matters worse, fossils exposed to the lower Mississippi River Basin's humid, semi-tropical climate by human activities or natural processes are either quickly weathered, eroded away, or hidden from view beneath a blanket of all-consuming vegetation.

This paucity of fossils is better understood by briefly reviewing conditions that exist at the state's two most well-known fossil locations. More than one hundred species of invertebrate and vertebrate macrofossils could be collected at one time from the Eocene-age Moodys Branch Formation at Montgomery Landing, but construction work along the Red River resulted in their burial by rock debris and vegetation. Roadside cuts made while constructing I-49 in the 1980s through Natchitoches Parish resulted in great exposures of the fossil-rich Cane River Formation of Eocene age. Unfortunately, in the three decades since, a combination of weathering, vegetation growth, and overzealous collecting by weekend visitors has altered this site to one practically devoid of fossils.

Rather than be discouraged by the relative absence of macrofossil-laden outcrops in Louisiana, novice paleontologists should take heart, because foraminifera, or forams, are abundant. Most commonly pinhead or smaller in size, these nearly microscopic organisms first populated the oceans of the world more than 500 million years ago, and today they are often the most abundant shelled organisms in many marine environments. More than one-third of present-day ocean floors is covered with foraminiferal ooze, a layer of mudlike sediment composed entirely of the multichambered shells of these single-celled animals. Forams feed on dissolved organic molecules, bacteria, and algae and, in turn, are eaten by worms, marine snails, sand dollars, and a variety of small fish. Most species are slow-moving bottom dwellers—a few burrow through seafloor sediment at speeds approaching 0.5 inch per hour—whereas others prefer an open-ocean lifestyle. Some exist for a mere few weeks, while others live for years.

Although forams are common in many sedimentary rocks, their extraction is a laborious process involving crushing a rock sample, washing it in a chemical solution, and wet screening it through a nest of very fine (0.25 to 0.5 millimeter) mesh sieves. A beginner's collection, however, can be easily gathered through a simple tried-and-true three-step process: fill a sandwich-sized ziplock bag with sediment collected in shallow water, air-dry the contents, and examine them under a 50-power binocular microscope. A teaspoon of sediment may well hold hundreds of foram shells. Accessible beaches are uncommon along the Louisiana coast, but shallow-water sediment samples are readily available along the 7-mile-stretch of Louisiana 1 on Grand Isle.

Approximately four thousand species of foraminifera exist today, and paleontologists recognize more than ten times that number of fossil species. Their miniscule size makes them unlikely candidates for most fossil collections, but anyone interested in the history of ancient life should at least be familiar with their existence, geologic range, and almost endless variety of forms.

Highly magnified images of a representative sampling of foraminifera fossils collected from Quaternary-age sediments of the Gulf of Mexico. Note that two species are listed as being agglutinated, *meaning their shell walls were built with minuscule sedimentary particles. Scale bars are 0.02 inch in length.* —Digital photomicrographs by Lorene E. Smith, Museum of Natural Science, Louisiana State University, Baton Rouge. With the exception of *Laevipeneroplis,* specimens are from the H.V. Howe collection of the LSU Museum of Natural Science. *Laevipeneroplis proteus* is from the personal collection of Barun K. Sen Gupta, Department of Geology and Geophysics, Louisiana State University.

37. State House, Maine

44° 18′ 26″ North, 69° 46′ 54″ West

A place to walk on remnants of the Iapetus Ocean.

For centuries sedimentary rock embedded with fossils has been used to enhance the exteriors and interiors of hotels, banks, and a wide variety of governmental buildings throughout the world. "Fossil Limestone," the trade name of an easily carved version of calcium carbonate, was the stone of choice of architects who designed the classical buildings that define the opulence of the Byzantium and Ottoman Empires. Soft when quarried and strengthened as its water content evaporates, its oyster-shell complexion embellishes the walls of the sixth-century basilica of Hagia Sophia and the sixteenth-century mosque of Suleiman the Magnificent, in Istanbul, Turkey.

Washington, DC, is home to a treasure trove of monuments, buildings, lobbies, and galleries constructed of fossil-bearing sedimentary rock. The Kasota Limestone, near the entrance doors of the National Museum of the American Indian, is tattooed with trace fossils, spaghettilike designs reflective of organisms that burrowed through the muds of Ordovician seas. More than 190 species of fossils are found in the Rice Chex complexion of the Salem Limestone walls of the Department of Labor building and the National Museum of Women in the Arts. Commercially known as "Indiana Limestone," this Mississippian-age rock was in vogue as an architectural stone throughout the United States during the late nineteenth and early twentieth centuries. Last, but most certainly not least, molds of clam-like pelecypods and tapered and spiraled gastropods impregnate slabs of 105-million-year-old "Cordova Shell Limestone" that accentuate the entrance to the dinosaur wing of the National Museum of Natural History.

Up north in the state of Maine, where the fossil record is relatively sparse, an unusual display of ancient life is preserved within the floor tiles of the state's capitol on State Street in Augusta. Constructed in 1832, the original Maine State House was enlarged over a period of two decades, between 1890 and 1910, using architectural stone marketed under the trade names "Radio Black" and "Champlain Black." Records identifying the source of this rock have been lost, but analysis suggests it was quarried in Vermont from the Ordovician-age Crown Point Formation, a carbonaceous limestone deposited in the Iapetus Ocean, the ancestor of the Atlantic. Four phyla of marine invertebrates have been identified, listed in a free fossil hunt guide available at the information desk next to the security station of the public entrance.

The most common fossils are pea-sized, Life Savers–shaped columnals, the broken remnants of crinoid stalks. Commonly called "sea lilies," these organisms anchored themselves to the seafloor and filtered food from the water using branching arms. Numerous columnals, accentuated by a central cavity, are readily visible through a hands-and-knees floor inspection. Shells of *Maclurites*, an extinct Ordovician gastropod, are easy to identify by their distinctive coiled shape. Two categories of coelenterates, the phylum that includes corals and jellyfish, are present: the solitary coral *Lambeophyllum*, and the extinct brussels sprout–shaped colonial organism *Stromatocerium*. The hardest fossils to find are bryozoans, otherwise known as moss animals. Shaped somewhat like a minute natural sponge, examples varying in size from 1 to more than 2 inches are embedded in tiles on the third floor. Other fossils are present, but they have not yet been fully identified due to their fragmented nature.

Paleontologists often jest that fossils are sure to be found in two places: the field and museums. In the northern corner of New England, however, a third locale is available for public review: in the tiled interior of the Maine State House, where legislation has been vigorously debated for more than 180 years and 475-million-year-old life is beautifully preserved under everyone's footsteps.

Scuffed, scratched, and waxed, the black limestone floors of the Main State House offer the best examples of readily accessible fossils in the state of Maine, such as this 3-inch-long crinoid stalk. Pen for scale.

This coiled invertebrate, Maclurites magnus, *was a resident of the Iapetus Ocean. Pen for scale.*

Thousands of bryozoan species exist today, but during Ordovician time these moss animals were the new residents of the world's oceans. Pen for scale.

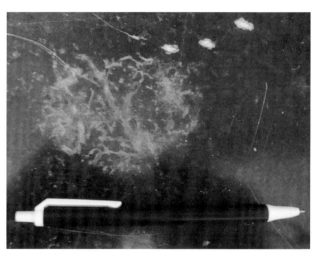

Is this a fossil, smudgy footprint, or splatter of dirt? For answers, stop at the information desk for a free copy of "Maine State House Fossil Hunt." Pen for scale.

38. Dinosaur Park, Maryland

39° 04' 15" North, 76° 52' 07" West

A toothy discovery in an open-pit rubble pile.

Prior to the late-nineteenth-century discovery of massive high-quality ore deposits in the Mesabi, Cuyuna, and Vermilion Ranges of northeastern Minnesota, the industrial vitality of America was largely dependent on ores of secondary quality mined from scattered, small-to-middling-sized iron districts. New England depended on limonite ore from the Salisbury District of Connecticut, the Robesonia District of eastern Pennsylvania first produced magnetite ore around the end of the eighteenth century, and Ohio's economic status grew when the Hanging Rock Iron Region produced its first ton of hematite in 1818.

In 1830 the eastern seaboard of Maryland was a region on the move. Overland travel routes were well established, and there was talk that a railroad might soon cross the state. The region was strategically bookended by the nation's capital to the south and the booming city of Baltimore to the north. Life was good, and bragging rights were accentuated by the discovery of ore-bearing clay in exposures of the Arundel Formation.

Arundel iron ore, a type of bog ore, is a soft and spongy siderite deposit that precipitated from iron-bearing groundwater in sedimentary rock in marshes and swamps. Lumpy masses of siderite, some stagecoach in size, were dug from the Cretaceous-age bedrock, broken into smaller pieces, loaded onto wagons, and hauled to regional iron furnaces and rail depots. During the boom years that ended around 1855, dozens of mines operated in the 5-mile-wide ore zone that extended from the northeastern corner of Maryland south into Virginia. Thirty years later, however, many operations had closed and the industry was all but history. Interest in the Arundel Formation continued, however, focused on a second resource known to but a few paleontologists. Fragmented dinosaur bones, present but hardly ubiquitous, are scattered throughout the layers of clay like maraschino cherries in a Christmas fruitcake.

In 1858, while conducting fieldwork that would lead to the first geological map of Maryland, Philip Tyson, the state agricultural geologist, received two cone-shaped teeth that had been found in the rubble of an open-pit iron mine near the village of Muirkirk. Tyson asked Christopher Johnston, a local doctor who also practiced dentistry, to examine the teeth. Finding a star pattern in a cross-section slice of one specimen, Johnston called it *Astrodon*, meaning "star tooth." Seven years later, Joseph Leidy, renowned as the father of American invertebrate paleontology, formally announced the tooth as evidence of a new dinosaur species, *Astrodon johnstoni*, the first sauropod described in North America.

This four-legged sauropod was 60 feet long nose to tail, 30 feet tall, and weighed 20 tons. *Astrodon* strolled across the Maryland landscape feeding on ferns, conifers, ginkgo branches, and other Early Cretaceous plants. In 1998 *Astrodon* was upgraded from Maryland's first known dinosaur to its official state dinosaur.

Most of the early mining locations are covered by urban development, but one small plot is still accessible. Located 1 mile north of the Muirkirk discovery site, Dinosaur Park, in the 13200 block of Mid-Atlantic Boulevard in Laurel, is advertised as the most important dinosaur fossil site east of the Mississippi River. Twice a month, its fenced-in mound of Arundel clay is opened to the public for the discovery of 115-million-year-old fossils of dinosaurs, turtles, crocodiles, and the occasional early flowering plant. Fossils found are kept in the park, however, so others can learn from them and because it is not unusual to find pieces that fit together to make a whole specimen. Small in acreage, Dinosaur Park is big in its representation of the history of the Arundel mining district and an earlier time when "terrible lizards" roamed the land.

The size and shape of this 2-foot-long Astrodon *jaw was an important clue in determining the size and shape of the dinosaur's head.*
—Courtesy of the Maryland-National Capital Park and Planning Commission

Artistic interpretation of Astrodon, *based on the size and shape of key leg, neck, and tail bones (shown in brown).* —Courtesy of the Maryland-National Capital Park and Planning Commission

The thrill of discovery— unearthing a weathered leg bone from its 115-million-year-old grave site at Dinosaur Park. —Courtesy of the Maryland-National Capital Park and Planning Commission

The dinosaur teeth found embedded in Arundel Formation rock are typically several inches long. This broken specimen endured the deteriorating effects of burial and time rather well. —Courtesy of the Maryland-National Capital Park and Planning Commission

39. Sandy Mile Road, Maryland

39° 42′ 44″ North, 78° 13′ 47″ West

A snowdrift of white sand is reminiscent of words Plato wrote more than 2,000 years ago.

Buried within the text of Plato's *Phaedrus*, composed between 375 and 365 BC, is a familiar axiom: "Things are not always what they seem; the first appearance deceives many." The central meaning of Plato's declaration is just as applicable today as it was more than 2,000 years ago. For example, when examining fossil-bearing rock in a roadcut or mountainside outcrop, many a field paleontologist has discovered that things are most certainly not always as they at first seem. An educational illustration can be seen at a roadside exposure of Oriskany Sandstone located within the shadow of I-68, 2.4 miles west of downtown Hancock, along Creek Road and then 0.1 mile south on Sandy Mile Road. Here in a snowdrift of white sand, examples of actual fossil shells appear to be common, but close inspection proves otherwise.

Proto–North America 400 million years ago, seated in the southern hemisphere, basked within an arid tropical environment similar to that of the modern-day Persian Gulf. Sediments eroding off the youthful Appalachian Mountains were deposited along the east coast of the Tippecanoe Sea and subjected to prolonged high-energy wind, currents, waves, and tides. Over time, mechanical and chemical forces weathered the less stable minerals of these sediments, leaving a 150-foot-thick blanket of 98-percent-quartz sandstone that today is ridge-forming rock extending from Virginia north into Pennsylvania. This Oriskany Sandstone sand was extensively mined for use in glass manufacturing, and it was even used to make lenses for the Hubble Space Telescope.

In western Maryland, the Oriskany contains two major fossil groups: crinoids and large, thick-shelled brachiopods. Crinoids are seldom found, the supposition being that the remains of the animal were broken apart after death and either dissolved or were carried offshore into deeper waters. On the other hand, evidence of brachiopod shells does remain. The shells face convex up, an orientation indicative of postmortem burial in an agitated environment, as this is the most stable position for any saucer-shaped mass in high-energy conditions.

At the Sandy Mile Road outcrop, however, actual brachiopod shells are not found. Rather their fossil evidence remains in the form of molds and casts, artifacts created in the surrounding rock material by the exterior or interior of a fossil shell. An external mold is an impression in rock showing the markings of the outer shell, whereas an internal mold shows the internal surfaces of the shell. A cast is created when rock material fills in a natural mold left by a shell. Three distinctive genera of brachiopod molds and casts are found within the outcrop: coarsely ribbed *Costispirifer*, the beautifully scalloped *Meristella*, and *Acrospirifer*.

Roughly 500 yards southeast of the Oriskany roadcut, the crumbly sandstone changes abruptly to compacted exposures of the Keyser Formation. Deposited within shallow water and tidal flats in Early Devonian time, this bluish, cherty limestone contains an array of small brachiopods, bryozoans, and coral fragments distributed among the weathered blocks and talus heaps of an abandoned quarry. Brachiopod genera that can be collected here include *Strophonella*, *Stropheodonta*, *Atrypa*, and *Rhipidomella*.

This site has great educational value, because fossils of a decidedly different nature lie within a stone's throw of each other. The porous Oriskany Sandstone was chemically altered, a process that dissolved the original fossils and transposed their likenesses as trace fossils. In contrast, the Keyser Formation fossils endured the years since their demise with minimal change and today look almost exactly as they did 400 million years ago.

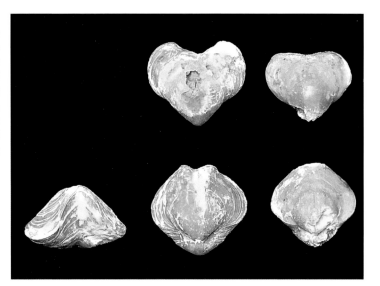

The brachiopod genus Meristella, *a Devonian-age index fossil found in the form of molds and casts in the Oriskany Sandstone, is identified by its smooth, oval shell (top right) and prominent incurved beak (bottom middle). These specimens, from Oklahoma, document* Meristella *in unaltered form. Each specimen is about 1 inch wide.* —Courtesy of Jayson Kowinsky

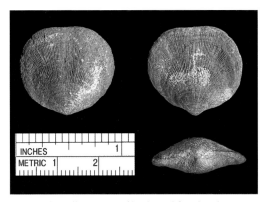

Rhipidomella, *a genus of brachiopod found in the Keyser limestone, has been found on every continent except Antarctica. Ruler for scale.* —Courtesy of Jayson Kowinsky

Fossils that are properly unearthed from Keyser Formation limestone and cleaned can be museum quality, such as these specimens of Atrypa, *an extinct brachiopod recognized by its distinctive growth lines. Ruler for scale.* —Courtesy of Jayson Kowinsky

The outer 0.5 inch or so of the Oriskany Sandstone has differentially weathered to a dark-gray color, leaving an irregular, pitted surface in which molds and casts of complete and broken shells can be seen. Pen for scale.

40. Beneski Museum of Natural History, Massachusetts

42° 22′ 19″ North, 72° 30′ 52″ West

A plethora of slabs preserve imprints of a long-lost world.

Among those intrigued by the world of fossils, possibly no other topic stirs the heart and the mind as much as dinosaurs, precursors to today's birds and masters of the Mesozoic Era food chain. Their bulk, geographic range, and ferociousness—or lack thereof—are topics that continue to stir controversy. After almost two centuries of study, even the question of how many there were cannot be answered through consensus, although those in the know generally acknowledge that there were about three hundred genera and seven hundred species. Evidence of these "terrible lizards" is found in many regions of the United States, but the Beneski Museum of Natural History, on the campus of Amherst College, stands out. Here, 1,700 slabs of rust-colored Triassic-age sandstone harbor a story of discovery, confusion, and eventual acceptance.

While plowing a field in western Massachusetts in 1802, twelve-year-old Pliny Moody overturned a nondescript stone slab to reveal digit-like impressions on the underside. Seeking advice from the village elders, he was told they were the "tracks of Noah's raven," a reference to the large, crow-like bird released by the biblical Noah from his ark and never seen again. While the word *dinosaur* would not be coined for another four decades, these "raven" tracks are believed to be the earliest dinosaur discovery in the United States.

In 1835 tracks of a similar nature were found in flagstones used to improve the streets of Greenfield, a village north of South Hadley. After several onlookers identified them as "turkey tracks," they were brought to the attention of Dr. James Deane, a resident naturalist and physician. Recognizing their scientific importance, he sought the advice of Edward Hitchcock, professor of geology at Amherst College.

Hitchcock studied both the Moody and Deane imprints and declared they were not those of either a raven or a turkey; rather, they belonged to giant "antediluvian" birds. One year later he published his interpretations and announced their study as the beginnings of a new science, today known as *ichnology*—the study of trace fossils, including fossil tracks. He spent his remaining years assembling the more than twenty-one thousand fossil tracks that constitute the core of the Hitchcock Ichnological Cabinet, the world's largest collection of dinosaur tracks. Professor Hitchcock died in 1864, unaware that many of his peers were beginning to recognize the Moody-Deane rock imprints as those of dinosaurs.

The Beneski Museum prints were quarried from the Deerfield Basin, one of some twenty or more sandstone-and-shale-filled half grabens created when the supercontinent Pangaea broke apart during the Triassic Period. The environment of the basin was similar to that of today's rift valleys of East Africa: semiarid and quenched now and then by monsoonal rains. Ephemeral lakes became the gathering sites for dinosaurs that left lasting prints in the shoreline mud while seeking food and water.

The Beneski tracks were assigned to a range—from seven to fourteen—of herbivore and carnivore species. The uncertainty has to do with the difficulty of distinguishing juvenile from mature imprints. Track morphology, however, is useful for differentiating herbivores from carnivores, calculating speed of motion—most walked slowly, but some could trot up to 30 miles per hour—and determining whether a tail was dragged or held erect. One slab, possibly the crown jewel of the exhibit, tells the tale of a small, water-sodden vertebrate plodding home, tail laid low, through a wind-driven Triassic rainstorm.

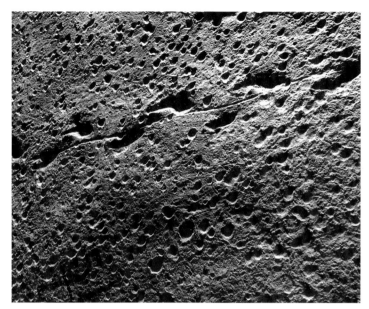

These 0.5-to-0.75-inch-wide footprints, tail-drag imprints, and unusually well-preserved raindrop impressions tell the soggy tale of a 200-million-year-old crocodile-like animal seeking shelter from a rainstorm. —Courtesy of the Beneski Museum of Natural History, Amherst College, the Trustees of Amherst College

Dinosaur tracks, complete with pad and claw impressions, are preserved as natural casts in this 8-inch-wide view of a slab originally used as a sidewalk in Middlefield, Connecticut. —Courtesy of the Beneski Museum of Natural History, Amherst College, the Trustees of Amherst College

Several types of dinosaur and invertebrate tracks are highlighted on this 5-foot-wide slab of brown shale from near Turners Falls, Massachusetts. —Courtesy of the Beneski Museum of Natural History, Amherst College, the Trustees of Amherst College

41. Horseshoe Harbor, Michigan

47° 28′ 24″ North, 87° 48′ 17″ West

Ancestral organisms found in a billion-year-old continental scar.

Any discussion of the geologic reconstruction of ancient geographies invariably involves the mention of two great supercontinents—Rodinia and Pangaea. The history of Pangaea has been extensively researched, from its birth 300 million years ago to its breakup beginning 100 million years later, when the push-and-shove tectonics responsible for its consolidation gave way to stretch-and-pull forces. The history of Rodinia is not as well understood. Assumed to have formed around 1.2 billion years ago, it existed for some 450 million years. Many details of its life span have been lost to the destructive forces of metamorphism and erosion, but evidence of its birth is found in the Midcontinent Rift System, a 1,000-mile-long arc of billion-year-old Precambrian rock buried in the upper Midwest. For 22 million years this ancestral counterpart to today's East African Rift System grew in intensity and width, vomiting lava that cooled as a 6-mile-thick column of volcanic rock later covered by an equally thick pile of sedimentary rock.

The environment of late Precambrian time was harsh, and barren, and characterized by rising oxygen levels and rampant erosion unchecked by the presence of soils and forests. Terrestrial life had yet to evolve when the Midcontinent Rift was experiencing its fifteen minutes of geologic glory, but bulbous fossils of aquatic life do exist within the rift rocks of Horseshoe Harbor, at the extreme tip of the Keweenaw Peninsula in Michigan.

The fossils have a swirl-like appearance and are interbedded with conglomerate, mudstone, and thin carbonate layers. Most paleontologists label them stromatolites, cyanobacteria that have been traced back 3.5 billion years to the earliest chapters of Earth. Formerly termed blue-green algae, these organisms exist in both unicellular and multicellular forms.

Present-day stromatolites resemble a cross between a layered rock and a cauliflower. The outer surface swarms with *cyanobacteria*, microorganisms characterized by the presence of chlorophyll and the liberation of free oxygen during photosynthesis. Cyanobacteria colonies secrete sticky masses of mucus that trap and bind together grains of sand, creating a mineral mat that reacts with dissolved calcium carbonate to form limestone. As more mats are produced, the stromatolites grow bigger and older, but the process is slow, developing approximately 1 inch of material every fifty years.

Modern stromatolites are commonly found in very salty lakes and marine lagoons, extreme conditions that protect the cyanobacteria from grazing animals. Analyses of the Horseshoe Harbor stromatolites and their host rocks, however, suggest the cyanobacteria lived in freshwater lakes dotting the barren and arid terrain of the Midcontinent Rift. This contrarian lifestyle raises an interesting question: How were these Precambrian rift lakes colonized with cyanobacteria when the nearest ocean was hundreds of miles away? Transport in the webbed feet of migrating ducks or on the scales of fish was not possible, as no such life-forms existed at the time.

Some paleontologists believe wind-borne transport was the solution to the dilemma; they consider the Horseshoe Harbor stromatolites as evidence of nature's early, albeit unsuccessful, attempt to cover the land with a mantle of flora and fauna. Others disagree. The history of these unusual fossil mounds remains a mystery involving some of the very earliest forms of life on Earth. Scientists may one day find consensus, but until then perhaps the best answer remains blowing in the wind.

A 3-foot-wide view of an assemblage of at least seven stromatolites. The light rock is the organic layers, and the brown rock the interlayered sediment. Canadian quarter (near center) for scale.

Cross section of a 3-foot-long Horseshoe Harbor stromatolite. These layered colonies of Precambrian cyanobacteria were some of the first forms of life to produce oxygen through photosynthesis.

Located to the left of this massive exposure of conglomerate at the west end of Horseshoe Harbor, the stromatolite colonies (not visible in this photograph) lie beyond a block-long thicket of shoulder-high vegetation.

42. Lafarge Fossil Park, Michigan

45° 04′ 53″ North, 83° 26′ 56″ West

A trio of extinct inhabitants offers up evidence of a long-lost inland sea.

Regardless of size, shape, weight, height, nationality, or gender, every individual who hopes to become a professional paleontologist must go on a first fossil field trip, a type of initiation. Once all the necessary equipment has been gathered, including ziplock bags, safety glasses, hammer, magnifying lens, and fossil field guide, protopaleontologists must address a litany of questions: Where do I go? What kind of rock should I look for? What kind of fossils might I hope to find? What can the fossils tell me of worlds lost to the passage of time? The latter is perhaps the most important question to ask.

For anyone traveling through the Lower Peninsula of the Wolverine State, the Lafarge Fossil Park and Besser Museum, headquartered at 491 Johnson Street, in Alpena, is an indoor-outdoor treasure trove of information and discovery worthy of that first field trip. The museum introduces the budding paleontologist to the Devonian Period of geologic time, and the adjacent open-dawn-to-dusk park contains an abundance of 400-million-year-old fossils that can be collected and taken home.

Constructed of eighty-three massive blocks of Alpena limestone mined from nearby quarries, the walls of the park surround more than 150 tons of crushed rock containing a rich inventory of fossils. Among the dozen or so different genera found, the horn corals, trilobites, and crinoids are especially interesting. Novice collectors value their eye-catching appearances, and professional paleontologists value them as indictors of prehistoric environments.

All three genera thrived in epicontinental seas that fluctuated across the embryonic landmasses of the Paleozoic Era, which lasted 289 million years. Horn corals, distant relatives of the jellyfish, lived either a solitary or colonial lifestyle, permanently anchored to the seafloor. Each straight or curved, cone-shaped, often wrinkled, exoskeleton housed a cluster of tentacles that gave the animal a flowerlike appearance when it filtered food from its warm-water habitat.

Trilobites were hard-shelled, segmented critters that meandered through a variety of marine environments and curled up like a modern-day roly-poly bug when threatened. They periodically molted, shedding of a series of exoskeletons as they grew larger. Their eating habits varied. Some were scavengers that vacuumed the seafloor, whereas others were free-swimming predators that favored plankton. More than twenty thousand species are known, perhaps making them the most diverse group of extinct organisms ever. They prowled their saltwater world guided by stereoscopic vision, a sense that paleontologists believe no other aquatic form of Paleozoic life possessed.

Devonian-age crinoids, a relative of starfish and sea urchins, preferred relatively clear, shallow to deep water that was home to massive volumes of minute organisms, a source of food they shared with their trilobite neighbors. Contemporary crinoids are commonly called "sea lilies," and like their extinct cousins, they attach themselves to the seafloor with a segmented stalk.

The fossils of Lafarge Fossil Park have helped scientists piece together the landscape of the Lower Peninsula of Michigan during Devonian time. Scavenging trilobites molted shells and crinoids fed among forests of waving coral tentacles in a shallow, inland ocean bathed by gentle breezes. Fossil-bearing strata are present throughout the world. Each specimen harbors a singular interpretation of long-lost environments. By assembling their individual narratives into a continuum, it becomes clear that life is ever evolving and Earth is ever changing.

The principal body part of the crinoid is a walnut-sized crown that typically has a star-shaped pentaradial symmetry—science-speak for five sections that are identical in construction and appearance and arrayed around a central axis. —Courtesy of Lafarge Fossil Park

Horn corals are found in two forms at Lafarge Fossil Park: as 2-to-3-inch-long specimens firmly entrenched in Alpena limestone, above (quarter for scale), and slightly smaller fossils that long ago weathered from their rocky tomb, below. —Courtesy of Lafarge Fossil Park

Besser Museum examples of wampum, threaded strings of 400-million-year-old crinoid columnals valued by Native American tribes throughout the eastern United States both as a form of currency and as stylish costume jewelry.

43. Petoskey State Park, Michigan

45° 24' 29" North, 84° 54' 09" West

A bit of elbow grease can turn an extinct coral into a rock star.

It should come as no surprise that when asked to define the personality of an honest-to-goodness fossil enthusiast, paleontologists would deliver a variety of answers differing in content and structure. That said, each reply would surely contain similar words, such as *interested*, *dedicated*, *determined*, and *inquiring*. Considering the overwhelming variety of fossils found in the world, perhaps the spirit of inquiry is one of the most essential personality traits for fossil enthusiasts. Whether amateur or professional, every aficionado of ancient life should constantly be seeking answers to questions both new and unsolved: What is the phylum of this specimen? How was this example fossilized? Is this genus colonial or solitary? Individuals curious about the fossilized bedrock of the Lower Peninsula of Michigan may ask one more—most unusual—question: When did the fossil become a stone? The answer is around 390 million years ago, during the tranquil years of the Devonian Period.

North America was then relatively young. Its mature state of tectonic development lay in wait far beyond the geologic horizon. The Kaskaskia Sea, warm and oxygenated, inundated the midwestern portion of the continent as far north as the Canadian border. Colonies of nematodes, earthworms, spiders, and centipedes overran an array of terrestrial habitats. Primitive forests shaded the expanding flanks of the developing Appalachian Mountains. Earth was experiencing its first green revolution. In the Kaskaskia Sea, the deep waters of the interior of the mitten-shaped Lower Peninsula of Michigan were separated from surrounding shoals by a circular band of reefs. The environment, not unlike that of today's Bahamas, was ideal for a variety of fauna, including trilobites, crinoids, bryozoans, brachiopods, and corals—especially corals. One coral in particular is of special interest to paleontologists, rock hounds, jewelry makers, and tourists: the ubiquitous and always entrancing *Hexagonaria percarinata*, the fossil that became a stone.

Hexagonaria percarinata is an extinct, colonial, seafloor-bound animal that lived in symbiotic relationship with an assortment of other marine life. Each individual soft-bodied polyp was housed within a theoretical six-sided *corallite*, an exoskeleton constructed of calcium carbonate. The colonies were composed of hundreds, if not thousands, of corallites linked together to form an elaborate hexagonal structure. Specimens of *Hexagonaria percarinata* generally occur as golf-ball-to-basketball-sized fragments that glaciers eroded from limestone bedrock during the Pleistocene Epoch. In the "raw," they appear as ordinary chunks of limestone of varying shades of gray, but their diffuse beauty is magically revealed when the fossils are immersed in water. With patience and polish, these uninspiring dollops of fossil coral can be transformed into gem-quality stones. In fact, polished versions called Petoskey stones are Michigan's official state stone.

Hexagonaria percarinata fragments are found along the 50-mile stretch of Lake Michigan shoreline extending from Traverse City to Petoskey. Many fine specimens have been discovered along the shallow shoreline of Petoskey State Park, advertised as the home of the Petoskey stone. Searches in the early months of the year, after winter storms have churned up offshore accumulations of limestone gravel eroded from Alpena limestone bedrock, tend to be most fruitful.

To transform a fossil into a Petoskey stone, file the specimen to the desired shape. Then, rub it with wet 220-, 400-, and 600-grit sandpaper in that order, a process that highlights a hexagon pattern resembling the rays of the sun. Finally, buff the find with mineral oil and a velvet cloth—then and only then does the fossil become a Petoskey stone. If after going through these steps the finished product is still lackluster, remember, practice makes perfect.

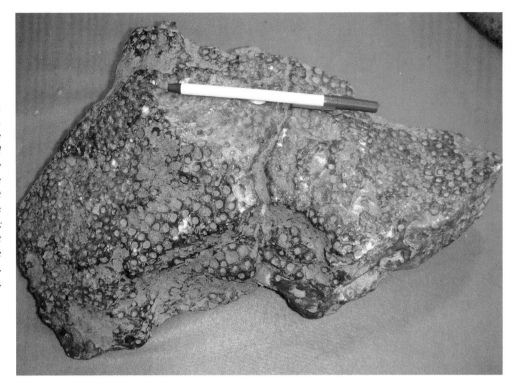

Hexagonaria percarinata *specimens are often found with naturally occurring crude oil. The amount of oil determines the fossil's color: light brown if not much, and dark brown if a lot. Pen for scale.*

This dry, natural specimen of Hexagonaria percarinata *cannot be labeled a Petoskey stone until it is polished to perfection. Pen for scale.*

Both paleontologists and gemologists value highly polished Petoskey stones. In 1965 this remnant of early Paleozoic time became Michigan's official state stone. Pen for scale.

44. Gunflint Chert, Minnesota

48° 05′ 31″ North, 90° 49′ 31″ West

Rocks that provided long-sought evidence of the dawn of life.

Prior to the 1950s, a description of the fossil record of Precambrian time could be presented with a paragraph or two. The flippant response to a university-level test question on the topic might read, "Life did not exist during Precambrian time." When asked to lend their names and reputations to published statements on the subject, geologists and paleontologists were often reticent. Even Charles Darwin labored over the question of Precambrian life, writing in 1859, in *On the Origin of Species*, "To the question why we do not find rich fossiliferous deposits belonging to these assumed earliest periods prior to the Cambrian system, I can give no satisfactory answer … The case at present must remain inexplicable."

For more than one hundred years following Darwin's declaration, members of the paleontological community debated the existence of fossils older than 500 to 600 million years. Many investigators considered the Precambrian a biologically barren time. A typical mid-twentieth-century historical geology textbook addressed the topic in words that echoed Darwin's exasperation: "As the discovery of Pre-Cambrian fossils has been a major objective among paleontologists for many years, intensive searches have been carried on wherever there seemed to be any possibility of finding such organisms. The results have been so meager that the task seems almost hopeless" (Hussey 1947).

Then, as with other abrupt turning points in the history of science, thinking was turned on its head. In 1954 fossil microbes associated with layered stromatolites were discovered in the 1,878-million-year-old Gunflint Chert, an iron-bearing rock that straddles the Minnesota-Ontario border. Microscopic studies revealed spore-like spheres, rods, and hairlike filaments displaying bilateral symmetry and supposed cellular differentiation.

Announcement of the Gunflint discovery spawned widespread surprise, denial, and disbelief nurtured by thoughts that such a find would not only more than triple the time frame of the history of life on Earth—an unacceptable leap of faith for many paleontologists—but would also upset long-standing scientific dogma. The size of the newfound bacterial and algae fossils was another deterrent to acceptability: estimates indicated that millions of specimens could be embedded in 1 cubic inch of rock; thus, they were too small for anyone to see or possibly even imagine.

Within a decade, however, scientists had documented their authenticity as the oldest fossils in the world, assigning them generic names such as *Kakabekia*, *Gunflintia*, and *Eosphaera*. Considered the earliest photosynthetic organisms, they played a leadership role in altering Earth's biosphere; they added oxygen to an oxygen-free atmosphere, paving the way for the evolution of organisms that use oxygen for energy. The discovery of the Gunflint fossils stimulated a worldwide search for additional evidence of Precambrian life. Two 3.5-billion-year-old rocks, one in western Australia, the other in South Africa, share the honor of harboring the world's oldest fossils: fossilized blue-green algae structures, or stromatolites.

Photograph-only exposures of glacially polished stromatolites of the Gunflint Chert occur a short hiking distance from the parking lot of the Magnetic Rock Trail, 1.25 miles north of the intersection of County Highway 12 and US Forest Service Road 1347, north of Grand Marais. Paleontologists and biologists alike associate the discovery of these beds of wavy rock with the birth of Precambrian paleobiology, the academic discipline that endeavors to unravel the enigmas associated with the dawning of life on Earth.

High-magnification thin section of Gunflintia minuta *from the Gunflint Chert, showing a diverse collection of individual filaments of blue-green algae that measure about 0.00004 inch wide.* —Courtesy of Gene LaBerge, University of Wisconsin, Oshkosh

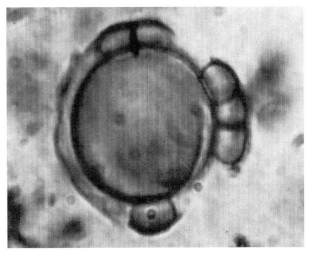

High-magnification thin section of Eosphaera tyleri, *a Gunflint Chert microfossil of unknown affinity, although it has characteristics similar to modern-day red algae. The diameter of this specimen is less than that of a human hair.* —Courtesy of Gene LaBerge, University of Wisconsin, Oshkosh

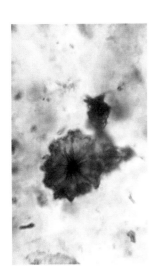

High-magnification thin section of Kakabekia umbellata, *a 0.001-inch-diameter, parachute-shaped Gunflint Chert microfossil with an uncertain biologic affinity.* —Courtesy of Gene LaBerge, University of Wisconsin, Oshkosh

A glacially polished cross section of stromatolite structures in the Gunflint Chert. Paleontologists believe this diverse group of microscopic Precambrian organisms thrived as microbial mats. Twig of blueberries for scale. —Courtesy of Mike Jirsa, Minnesota Geological Survey

45. Wangs, Minnesota

44° 24' 39" North, 92° 59' 02" West

Minnesota's best formation for collecting Ordovician-age fossils.

Since it became a state in 1858, Minnesota has acquired a handful of nicknames: Gopher State, Land of 10,000 Lakes, North Star State, Vikings State, Bread and Butter State, State of Hockey, and New England of the West. Seven monikers are enough for most people, but some paleontologists might insist on an eighth: Land Somewhat Lacking in Fossils.

Minnesota's geologic history dates back to the near-dawn of the Archean Eon, 3.5 billion years ago. More than 60 percent of the area of its basement terrain is constructed of bands of igneous and metamorphic rock that form the Canadian Shield, the ancestral nucleus of the North American continent. Fossils are rarely found in these rocks. The intense temperature and pressure conditions in which they were created guaranteed the absolute destruction of evidence of preexisting life. Almost all fossils, however, are preserved through the process of sedimentation: before decay and consumption by predators can take place, they are preserved in weathered or precipitated material that is later compressed into sedimentary rock. Exposures of Paleozoic-age sedimentary rock, found exclusively in southeastern Minnesota within a triangular area defined by the Iowa border, the Mississippi River, and I-35, contain a representative sampling of Gopher State fossils.

During Ordovician time, Minnesota straddled the equator and was partially inundated by a shallow, warm sea. The climate was tropical and marine life abounded. Sequences of shale and limestone deposited in this utopian environment have attracted generations of weekend and professional paleontologists intent on discovering a collection-defining specimen. The formation of choice is the Decorah Shale, a greenish-gray rock of Middle Ordovician age that harbors four different fossil types: horn corals, brachiopods, bryozoans, and crinoids.

Horn corals are the calcareous remains of extinct, solitary, bottom-attached animals that captured floating sea life with flowerlike tentacles. Brachiopods are confined within a shell composed of two unequal halves. Like horn corals were, they are filter feeders, but their body parts remain inside their shell.

Bryozoan species were both abundant and diverse during Ordovician time. Two contrasting examples of these small—tinier than a typical pinhead—colonial marine animals are found in the Decorah Shale. Amateur collectors often misidentify the broken twiglike specimens of *Hallopora* and *Batostoma* as the bones of small, rodent-like animals. When compared to the spongy texture of vertebrate bone, however, the cross sections of weathered specimens reveal a hollow and concentric form, the defining characteristics of bryozoan colonies. The unconventional shape of the second type, *Prasopora conoidea*, domed on top and flat along the base, is the reason they are often referred to as "gumdrop" and "chocolate drop" fossils. Specimens of this highly prized fossil typically have a diameter in the range of a penny to a half-dollar.

Crinoid fossils are common in the Decorah Shale, but discovering their small—button-sized or smaller—disarticulated Life Savers–shaped plates requires hands-and-knees outcrop exploration. In life these calcareous disks were arranged in a bead-like series to form a flexible stem connecting the crinoid root to its main body. Native American tribes strung crinoid disks together as decorative necklaces, but today these disks are the despair of field geologists because they have little practical value in age dating their host rock.

Roadside exposures of 460-million-year-old Decorah Shale can be found in many southeastern Minnesota locations. A favored and recommended site occurs along the east side of Minnesota 56, 0.3 mile north of the intersection with County Road 9 in the village of Wangs, between Rochester and Minneapolis.

Partially covered by vegetation, the weathered Decorah Shale exposure along Minnesota 56 immediately north of Wangs is home to a variety of 450-million-year-old invertebrate fossils.

Crinoids from Wangs resemble Life Savers, though their size and shape do vary, unlike the candy. Three examples of a partial flexible stem are left of the coin. Penny for scale.

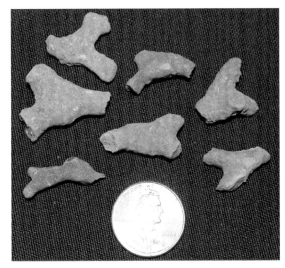

Colonies of Batostoma bryozoans grew in treelike fashion, but due to their fragile nature they are commonly found as fragments that resemble broken twigs. Penny for scale.

Prasopora conoidea bryozoans are conventionally termed "gumdrop" bryozoans for obvious reasons. After achieving worldwide distribution during Ordovician time, they became extinct ten to twenty million years later. Penny for scale.

46. W. M. Browning Cretaceous Fossil Park, Mississippi

34° 35′ 09″ North, 88° 37′ 56″ West

Fossil shark teeth for the picking at Mississippi's only geological park.

Jurassic Period paleogeography can be traced back to the time when the colossal supercontinent Pangaea began to come apart at the seams. Africa separated from South America, giving birth to the ancestral Atlantic Ocean, and then Australia and Antarctica assumed continental status. To the north Eurasia disassociated from North America, leaving Greenland, the world's largest island, in its wake. This convoluted process of landmass motion is still underway; satellite analysis indicates the distance between New York City and London increases roughly 1.25 inches per year, a length comparable to the annual growth of the human fingernail.

By Cretaceous time global sea level was at a high, the bulk of North America was located between 30 and 60 degrees north latitude, and it was approaching its current position on the globe. The ancestral Rocky Mountains were beginning to rise toward topographic prominence. In the east the developing Gulf of Mexico was modified by the Mississippi Embayment, a low-lying, fingerlike feature that extended hundreds of miles north along the valley of the ancestral Mississippi River. By Late Cretaceous time, rivers flowing off the northwestern slopes of the Appalachian Mountains were depositing loads of sediment into the 100-foot-depths of the embayment.

In 1990, while constructing a bypass to old US 45 in Prentiss County, workers unearthed two fossil-bearing beds that had formed in the embayment. From this stash paleontologists expanded their knowledge of northern Mississippi 75 million years ago, during Late Cretaceous time. Five years later this site was designated the W. M. Browning Cretaceous Fossil Park and opened to the public. It is located 0.6 mile north of Frankstown, at the intersection of County Road 7450 and Twenty Mile Creek and marked by a 5-foot-tall granite monolith.

Riverbank exposures in Browning park consist of Coffee Sand overlain by units of the Demopolis Formation, which comprises the 2-foot-thick, shark-tooth-bearing Frankstown Sand topped by an equally thick unnamed oyster bed. Clusters of boulder-sized concretions clog the valley of Twenty Mile Creek. Their history is the subject of debate, but one hypothesis is that sediment coalesced around a catalyst of shell fragments, similar to the way a pearl forms. Fossils from both terrestrial and marine environments constitute the inventory of the park: ammonites, bivalves, and petrified wood, as well as the bones of fish, turtles, and several species of dinosaurs. The most common specimens collected from the oyster bed are *Exogyra ponderosa*, extinct, thick-shelled, and stumpy oysters nicknamed "devil's toenails" by mollusk enthusiasts.

More than ten different types of shark teeth have been found in the Frankstown Sand, but only two genera are common. Both are middle to late Cretaceous index fossils. *Squalicorax* was a coastal predator that dined on a variety of fish and turtles and any terrestrial animal innocent enough to wade in shallow water—as evidenced by its tooth imprint on the foot bone of a duck-billed dinosaur found at another site. Flat and triangular with large roots, these teeth are the only Cretaceous ones known to have serrations. The genus *Scapanorhynchus* is distinguished by an elongated snout that supposedly housed electroreceptive sensors. This feature, along with a large tail fin, indicates this predator haunted the darkened deepwater domain. Both predators prowled the embayment.

The best time to visit Browning park is after an extended period of rainfall, when fossils are flushed from the stream bank. Bring a sieve with quarter-inch wire mesh, and be sure to sample deposits of coarse sediment as well as the deeper-water zones around and under the concretions.

Although boulder-like in appearance, the massive concretions that line Twenty Mile Creek were not rounded by stream action; rather, they formed within the adjacent formations as sediment cemented itself around a nucleus. —Courtesy of Jim Lacefield

Exogyra ponderosa *possessed a thick shell distinguished by ragged ridges—the greater the number, the older the oyster. Quarter for scale.* —Courtesy of Jim Lacefield

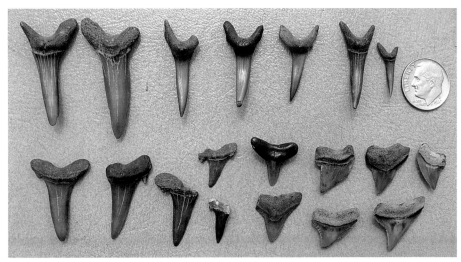

The teeth of two genera are common at the park: the dagger-shaped teeth of the "goblin" shark Scapanorhynchus *(upper row and bottom left) and the serrated, triangular teeth of* Squalicorax, *the crow shark (bottom right). Dime for scale.* —Courtesy of Jim Lacefield

47. Mastodon State Historic Site, Missouri

38° 22' 51" North, 90° 22' 52" West

A 13,000-year-old mastodon kill site in metropolitan suburbia.

The drive south from St. Louis along I-55 is similar to that on any interstate into and out of a major American city: the traffic is heavy, the speeds are fast, and the roadside signage beckons travelers to a myriad of fast-food outlets, gas stations, and miscellaneous related facilities. However, 2.5 miles south of exit 190, 425 acres of forest tell an intriguing tale of evolution and migration, early American culture, unexpected discovery, and scientific excavation that can break up the monotony. The Mastodon State Historic Site offers the opportunity to return to the final days of the great ice age. Here is the brief version of its tale, told in four chapters.

Chapter 1: Evolution and Migration. About 45 million years ago a population of elephant-like mammals evolved into a form recognized today as the mastodon, *Mammut americanum*. Seeking new territory, these 10-foot-tall, 5-ton shaggy beasts migrated through Eurasia and eventually reached North America by way of the Bering Land Bridge. Feeding on a diet of coarse herbaceous vegetation, they grazed and browsed their way as far south as central Mexico and then became extinct 10,000 years ago, victims of climate change or overhunting by resident populations, or possibly both.

Chapter 2: Early American Culture. While perhaps not members of the first wave of human migrants to the New World, the Clovis culture was the first well-established form of humanity to inhabit North America. Until recently, archaeologists believed this culture reigned for 4,000 years, starting around 14,000 years ago. Recent carbon-14 age dating has refined this span to a 300-year period, from 13,200 to 12,900 years ago. The presence of Clovis points determines if an archaeological site should be certified as belonging to this culture. These elongated, fluted projectile spear points were finely crafted from flint, chert, and obsidian. Clovis people shaped them to have slightly convex sides and a shallow, concave base. Archaeologists posit that these points were attached to wooden shafts used to hunt bison, horses, camels, and mastodons.

Chapter 3: Unexpected Discovery. Curious about the rumor of bones weathering from the banks of Rock Creek near the frontier community of Kimmswick, 1.5 miles southeast of the Mastodon State Historic Site, Albert Koch conducted some excavations in 1839 and assembled the collected fossils into a less-than-accurate skeleton he christened *Missourium kochii*, better known as the "Missouri leviathan." Following the exhibition of this believe-it-or-not discovery throughout the United States and abroad, it was correctly identified as the remains of a mastodon.

Chapter 4: Scientific Excavation. Mastodon kill sites are few and far between in North America. The first discovery was a bone spear point embedded in the ribs of a mastodon found in Washington State in 1977. Two years later, an artistically formed stone spearpoint was found with mastodon bones at the aforementioned Kimmswick bone beds. The style of the point was Clovis, making this discovery the first undisputed documentation that Clovis people hunted mastodons in North America.

With the 1979 Kimmswick find, the story of mastodons and their association with the Clovis people was completed. The Mastodon State Historic Site (1050 Charles J. Becker Drive) in Imperial (just northwest of Kimmswick) can provide you with everything you would wish to know, and more, about this tale of paleontology and archaeology. Here you'll find a full-size mastodon skeleton, a gigantic Harlan's ground sloth, and a representative Clovis culture campsite.

As can be seen with this 5-inch-long example, the teeth of mastodons, with high crowns and sharp cone-shaped crests, were best suited for mashing and grinding leaves, woody plants, branches, and twigs.

Mastodon means "breast tooth," a reference to their teeth's resemblance to the human breast. This tooth may have blackened by absorbing minerals from the soil it was buried in.
—Courtesy of Paul Ruez

Novice paleontologists often confuse mammoths with mastodons. Although both shaggy brutes were similar in size and appearance, their teeth are quite different. Mammoth molars are characterized by a variable number of closed, wear-resistant enamel loops that form the chewing surface. The flattened, cheese grater–like appearance of mammoth teeth suggests this Pleistocene-age vertebrate grazed principally on grasses, sedges, and rushes. Mammoths had the biggest grinding teeth of the animal kingdom. Quarter for scale.
—Courtesy of Paul Ruez

48. Riverbluff Cave, Missouri

37° 06′ 24″ North, 93° 19′ 43″ West

A national terrorist attack unexpectedly saves an unusual fossil treasure.

Just about every American who was alive in 2001 knows that 9/11 represents the month and day when nineteen hijackers crashed four commercial airplanes into the twin towers of the World Trade Center in New York City, the Pentagon in Washington, DC, and a field in rural Pennsylvania, killing 2,977 people. This story completely overshadowed all other news that infamous day, relegating it to inside pages or causing it to be dismissed altogether as unimportant. A case in point is an incident that took place in Springfield, Missouri.

A road-building crew had planned twelve dynamite detonations for the day. Two charges were in place when a nationwide ban on explosions took effect following the terrorist attack. Since they couldn't safely extract the charges, they were granted an exception to the order. The explosions blew open a massive cavity—2,000 feet long by 80 feet high—in Burlington Limestone, a rock very susceptible to karst erosion. A dank and musty odor emerged from the hole, and investigations later revealed that this 1.1-million-year-old subterranean grotto, now known as Riverbluff Cave, had been sealed some 55,000 years prior by rockfall.

For hundreds of thousands of years Riverbluff Cave housed an amazing menagerie of ice age animals: mammoth, horse, American lion, saber-toothed cat, armadillo, turtle, snake, earthworm, and at least one 1-inch-long millipede. It functioned as a refuge from weather and predators and a safe place to hibernate and give birth. A microscopic examination of a 1-foot-by-2-foot block of cave-floor sediment found as many as five thousand microfossils. The cave's limestone walls harbor a plethora of Mississippian-age echinoderms, the group that includes modern-day sea lilies and starfish.

The cave's claim to fame, however, are three ichnofossils, fossil traces left by organisms while moving, feeding, resting, or defecating. First, there are 2-foot-long claw marks 15 feet high on the cave walls, left by a giant short-faced bear. Considered the *T. rex* of the ice age and capable of running 40 miles per hour, this extinct bear generally stretched more than 10 feet tail to snout and was armed with a bite force of up to 2,000 pounds per square inch. Weighing on average 2,000 pounds, its heft alone was enough to scare other scavenging carnivores away from their kill, leaving it to dine in solitude. The height of the claw marks indicates they were left by a 12-foot-tall bear.

Second, the cave houses the largest concentration of flat-headed peccary hoof tracks in the world. Preserved in the cave floor, the excessive tracks are evidence that herds of these omnivore mammals used the site as a shelter. Standing 30 inches tall and weighing around 110 pounds, they became extinct 11,500 years ago. Lastly, there are coprolites, the ancient waste of former cave denizens. It is suspected that some of the fossil dung was left by a short-faced bear; if confirmed the fossil could provide details regarding the diet of these land predators.

Preparations are underway to open Riverbluff Cave to the public in the form of a continuing-research dig site. In the meantime, a well-organized collection of representative fossils is on view in the adjacent museum at 2327 West Farm Road 190. The tragedy of 9/11 will not soon be forgotten, but had it not happened all twelve dynamite charges would have been discharged that fateful day, and Riverbluff Cave—an ice age time capsule sealed for tens of thousands of years and now designated a national paleontology treasure—would have been destroyed.

Riverbluff Cave is adorned with a variety of speleothems: massive stalagmites project upward from the floor, and a multitude of soda straws, tubular stalactites that maintain the diameter of a drop of water and resemble a drinking straw, hang from the ceiling. —Courtesy of Kendra Mignard

Fully extended, this 1-inch-long segmented organism is believed to be the only millipede fossil dating to the Pleistocene Epoch ever found. Somewhere between 50,000 and 20,000 years old, it probably fed on decaying vegetation. Quarter for scale. —Courtesy of Kendra Mignard

These 2-foot-long five-digit claw marks belong to a giant short-faced bear, the largest bear, and perhaps the largest carnivorous land mammal, to have lived in North America. —Courtesy of Kendra Mignard

Fossil bone segments of a peccary foot. These ice age mammals were well adapted to frigid, dry, and dusty environments. Their descendants live today in Central and South America and in the US Southwest. Quarter for scale. —Courtesy of Kendra Mignard

49. Hell Creek Country, Montana

48° 00′ 35″ North, 106° 25′ 19″ West

***T. rex* and *Triceratops* discoveries in Hell Creek country.**

Tourists interested in river rafting adventures, slot canyon exploration, or rain forest trekking need not bother with northeastern Montana. However, for those seeking dinosaur finds, Hell Creek country is the place to be. Hell Creek Formation exposures in northeastern Montana, the type locality for the iconic *Tyrannosaurus rex*, have long captivated paleontologists. The remains of at least seventeen different species of dinosaurs have been extracted from the fresh- and brackish-water mudstone and sandstone strata of this Cretaceous-age formation. The genera *Tyrannosaurus* and *Triceratops* dominate the list, and the discovery and analyses of their bones are the reason why the Hell Creek area is arguably the most intensely studied dinosaur locale on Earth.

The first organized search for Hell Creek fossils was led by Barnum Brown, a largely self-taught Columbia University dropout who became known to generations of paleontologists as Mr. Bones. In August 1902 he made a brief, routine mention of a discovery in a letter to his supervisor at the American Museum of Natural History in New York: "Quarry No. 1 contains several bones of a large Carnivorous Dinosaur . . . I have never seen anything like it from the Cretaceous" (Dingus 2015). Through this matter-of-fact report Brown eventually became a celebrity as the discoverer of *Tyrannosaurus rex*, now known throughout the world as *T. rex* or the "tyrant lizard king."

Ninety-five years later, a *T. rex* skeleton that was 80 percent complete was unearthed from the same Hell Creek strata. Named after the nearby community of Fort Peck, three versions of "Peck's rex" are on display at the Fort Peck Interpretive Center: a fleshed-out, full-scale model, a cast of the skeleton, and a skull that housed cucumber-sized teeth capable of administering a bite force of 7,000 to 8,000 pounds per square inch. The 3-inch-long probable battle scars and signs of arthritis revealed by bone analyses suggest this animal led a tough life.

The so named "B. rex" specimen, an adolescent that died 68 million years ago and is catalogued as the oldest Hell Creek *T. rex* on record, made history in 2005. Scientists announced that they had found marrow-like tissue, common in female birds, on the inside surface of a thigh bone. Research involving this controversial material may answer questions about how dinosaurs evolved; how their muscles and blood vessels functioned; and if they were cold-blooded, warm-blooded, or both.

In 1965 workers with the Natural History Museum of Los Angeles unearthed the remains of a 9-foot-tall, 12-ton, 66-million-year-old trihorned *Triceratops* from Hell Creek strata. A replica of this specimen is on display at the Garfield County Museum in Jordan, Montana, along with *T. rex* teeth and skull and the domed skull of *Pachycephalosaurus*, a thrill for Harry Potter fans because it closely resembles the head of a dragon that rambles through the Hogwarts story. Believed second only to *T. rex* as the most belligerent dinosaur, *Triceratops* is the most commonly recovered dinosaur from the Hell Creek Formation.

Many paleontologists interpret the phrase "northeastern Montana" to mean "discovery locale of Cretaceous-age dinosaurs." Any exploration of this world-famous region's geology should begin in Jordan, where a sign on the south end of town says it all: "In the Heart of T-Rex Country." The exhibits at the Garfield County Museum and at the Fort Peck Interpretive Center, 93 miles to the north, tell the rest of the Hell Creek story.

A cluster of bones from the jaw of a juvenile, 14-year-old T. rex *lying in situ and upside down on a mud-cracked surface. Image is 7 inches across.* —Courtesy of Ron Giesler, ADVENTURE 360

The B. rex fossil site precariously balanced on a hillside exposure of Hell Creek Formation. B. rex is considered one of the most significant dinosaur discoveries because the specimen included the first dinosaur soft tissue and protein ever discovered. —Courtesy of Jack Horner, Museum of the Rockies, Bozeman, Montana

The brow horns (center) and lower jaw of a Triceratops *skull being removed from the Hell Creek Formation. The darker bones have been coated with a high-quality preservative that protects them from further weathering. Centimeter ruler for scale.* —Courtesy of Ron Giesler, ADVENTURE 360

50. Museum of the Rockies, Montana

45° 39′ 31″ North, 111° 02′ 42″ West

A museum to quench your dinomania.

English is the de facto language of diplomacy, the dominant language of the Internet, and a language characterized by continuing change. A recent listing of newly coined English terms included *fracking*, *selfie*, and *tweet*. This evolving means of communication invades every category of occupation, and paleontology is no exception. *Dinomania*, defined as a "growing interest in or enthusiasm for dinosaurs," is a relatively new word, but its meaning can be traced back to the middle years of the nineteenth century. In 1842 Richard Owen, the preeminent English anatomist of the day, examined bones and teeth that had been labeled *Megalosaurus* and coined the word *dinosaurian*, meaning "terrible lizard." Thirty years later, the opening shots of the bone wars (see sites 13 and 16), a duel-to-the-death conflict of ambitions, temperaments, and personalities, moved the global stage of vertebrate paleontology from the Old World to the New World. For a full quarter of a century Edward Cope and Othniel Marsh—dedicated fossil-finding antagonists—conducted an acrimonious war of words and deeds across the plains of the American West and, in the process, discovered more than one hundred new species of dinosaurs.

Then in 1902, Barnum Brown, a Kansas-born student of fossils better known to his peers as Mr. Bones, announced the discovery of *Tyrannosaurus rex*, or, as he put it, the "ultimate terrestrial predator." Seemingly overnight dinomania engulfed the ivied halls of universities and museums, and the epithet *T. rex* became a defining paleontological buzzword.

In 1980 the father-and-son team of Luis and Walter Alvarez announced that a nuclear winter, triggered by the impact of a rogue meteorite 66 million years ago, had caused the demise of the dinosaurs. It sparked a dinomania renaissance of sorts. Questions both intelligent and inane had to be addressed: Might volcanism have been the real cause of extinction? Did they die all at once or over a span of time? Is Nessie, the Loch Ness Monster, a survivor? If impact is the cause, where is the smoking-gun crater? Are hummingbirds really distant cousins of dinosaurs? With the 1993 release of *Jurassic Park*, a blockbuster movie depicting cloned dinosaurs breaking loose from their amusement park enclosures and running amuck on a Pacific island, dinomania peaked internationally. Theme-related books, comics, and video games flooded the children's market, and new fossil discoveries were announced in China, Argentina, Mongolia, South Dakota, and Africa.

Today family-friendly dinosaur exhibits are available globally. One of the very best is the Siebel Dinosaur Complex in the Museum of the Rockies (600 West Kagy Boulevard) in Bozeman, home to the largest collection of dinosaur remains in the United States. Displays large and small tell the story of the Age of Reptiles. "Montana's *T. rex*," one of the most complete specimens of *Tyrannosaurus* ever found, stands 12 feet tall and 40 feet long. The exhibit also showcases the largest *T. rex* skull ever unearthed, evidence of a burrowing dinosaur, and information about dinosaur behavior and skeletal *ontogeny*—that is, how organisms develop. For example, with age duck-billed dinosaur skulls changed shape and *Triceratops* horns reversed their orientation.

The *Jurassic Park* story line of a world of cloned reptiles remains a dream, but a visit to the Siebel Dinosaur Complex is the next best way to address a sudden attack of dinomania.

Unearthed in China, this pair of Cretaceous-age dinosaur eggs is similar in size and shape to those found in rocks of similar age in Montana. Quarter for scale. —Specimen from the collection of the Radford University Museum of the Earth Sciences

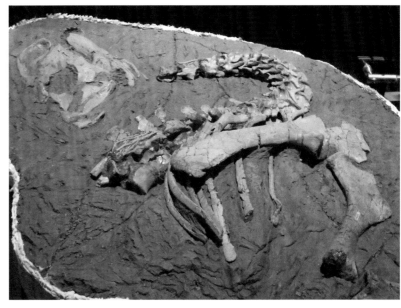

The necks and heads of some fossils are arched backward, such as with this herbivorous Tenontosaurus, *a posture attributed to severe muscle spasms the animals experienced while under predator attack and gasping for breath.* —Specimen from the Museum of the Rockies collection

Standing tall and in attack mode, the iconic T. rex that roamed Montana during the Cretaceous Period occupies a central position at the Museum of the Rockies. —Courtesy of the Museum of the Rockies, Bozeman, Montana

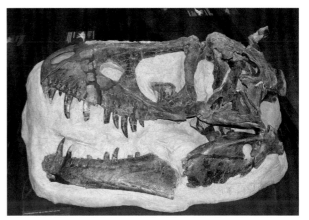

Daspletosaurus, *a close relative of* T. rex, *was a bipedal predator normally equipped with dozens of large, sharp teeth.* —Specimen from the Museum of the Rockies collection

51. Agate Fossil Beds National Monument, Nebraska

42° 25′ 33″ North, 103° 43′ 59″ West

A long-ago drought left evidence of dwarf rhinos and devil's corkscrews.

Millions of years ago the state of Nebraska was experiencing a multimillennial drying-out episode. The maturing Rocky Mountains interrupted moisture-laden air flowing out of the west, and the climate became cooler and increasingly arid. Animal populations adapted to new sources of food as grassland savannas, characterized by scattered watering holes, replaced the wetland forests of yesteryear. These critters recognized the Darwinian concept of adapt or perish.

All was well as long as rains dampened the land with regularity, but over time drought became the new norm. Once-seasonal streams became nonexistent and grasslands withered. With each passing season greater numbers of animals were drawn to the same shrinking oases, where they fought for a share of the diminishing vegetation and evaporating water. Their bones littered the landscape, today serving as fossil evidence of the vertebrate populations that inherited the Earth following the catastrophic demise of the dinosaurs. One such bone bed was discovered in 1885, 22 miles south of the high prairie community of Harrison. Twelve decades later, it and several other similar sites were amalgamated as Agate Fossil Beds National Monument.

Six vertebrate species dominate the roster of extinct animals. They ranged the fossil beds site 21 to 19 million years ago, during the early Miocene Epoch. A sampling of their nicknames—sloth foot, terrible pig, gazelle-like camel, and carnivorous bear dog—suggest a faunal diversity as fanciful and bizarre as that found today on the Serengeti Plain of Tanzania and Kenya. The unusual burrowing habit of *Palaeocastor* and the small size of *Menoceras* give special-interest status to two additional vertebrates.

Classified as a forerunner of present-day beavers, *Palaeocastor* was a land-loving rodent that lived inland, prairie-dog style, in well-drained ground, rather than underwater in a dam-built pond. Using their teeth, they dug 6-to-8-foot-deep spiral burrows—called *Daemonelix*, or "devil's corkscrews"—that bottomed out in subterranean nesting chambers. These vacated burrows filled in with compacted sediment that later weathered as corkscrew-shaped trace fossils. They were once thought to be the remains of a freshwater sponge or the taproots of a massive plant, until the controversy of their origin was laid to rest when a *Palaeocastor* skeleton was discovered inside one of the burrows.

More than two hundred catalogued specimens of *Menoceras*, a sheep-sized rhinoceros, have been unearthed from the sedimentary rocks at Agate Fossil Beds. Skull features allow paleontologists to identify the gender of specimens. A pair of bony lumps, the remains of side-by-side horns that were located at the tip of the animals' nose, identifies the males. This anatomical feature is in contrast to modern rhinos, which have nasal horns situated one behind the other. A skull with a smooth nasal bone is evidence of a female. Analyses of the extent of wear on the teeth of both genders indicate the males died at a younger age than the females, a mortality trait seen in today's rhinos—males often die from wounds inflicted during fights. It is believed that *Menoceras* populations equaled the estimated 30 to 60 million bison that dominated North America prior to the arrival of Europeans.

Interactive mammal exhibits are available daily at the visitor center, but the best fossil displays are perhaps found along the 1-mile-long Daemonelix Trail, along which in situ devil's corkscrews of the ancient *Palaeocastor* can be found at three locations.

This encased devil's corkscrew is prominently displayed at the halfway point of the Daemonelix Trail. —Courtesy of the National Park Service

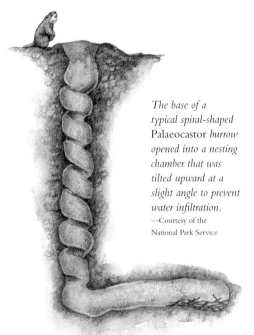

The base of a typical spiral-shaped Palaeocastor *burrow opened into a nesting chamber that was tilted upward at a slight angle to prevent water infiltration.*
—Courtesy of the National Park Service

About the size of a domesticated sheep, the fleet-footed Menoceras *roamed the Miocene plains of America.* —Courtesy of the National Park Service

During periods of drought a myriad of species died, leaving skeletal remains littering the area around evaporating water holes. This jumbled mix of casts of Menoceras *bones at the visitor center reconstructs a typical water-hole scene. Ruler for scale.*
—Courtesy of the National Park Service

52. Ashfall Fossil Beds State Historical Park, Nebraska

42° 25′ 13″ North, 98° 09′ 21″ West

A detailed and clear time-capsule panorama of the Miocene Epoch.

Like the eruption of Mount St. Helens in Washington in 1980, an event that killed fifty-seven people and hundreds of thousands of large and small animals, another northwestern US eruption was responsible for death across an extensive downwind swath of territory, but that is where the comparison ends. Bruneau-Jarbidge erupted 11.83 million years ago, one of eight supervolcanic centers that form the Snake River Plain, the desolate moonscape terrain of southern Idaho. Roughly 25 cubic miles of ash vomited into the atmosphere, one hundred times the volume discharged by Mount St. Helens and on a par with the 1815 eruption of Mount Tambora, the massive outburst associated with the following Year Without a Summer and the worst famine of the nineteenth century.

Around the northeastern Nebraska community of Royal, near the western limit of soils deposited by the ebb and flow of Pleistocene-age ice sheets, the land rises from the gentle slopes of the Dissected Till Plains to the more diverse and hilly Great Plains. This is tried-and-true Cornhusker country. It is also the land of the ubiquitous cow and coyote, occasional bison, melodic meadowlark, and iconic bald eagle.

A completely different animal population lived here in Miocene time. Packs of wild dogs, gregarious crowned cranes, herds of rhinoceroses, and the pedestrian secretary bird lie in the Ash Hollow Formation, a 1-to-6-foot-thick bed of gray volcanic ash composed entirely of microscopic shards of broken glass. This population seems like it belongs on the veldt of southern Africa, not the prairie of Nebraska. Chemical fingerprinting traces this ash back in distance and time to the Bruneau-Jarbidge eruption.

Immediately after the eruption, an ominous cloud of abrasive ash began its easterly jet-stream journey, blanketing the countryside. In Nebraska, a myriad of animals that were gathered about a favorite water hole ingested the powdery volcanic material while quenching their thirst. The smaller animals died first, by asphyxiation, as the shards of glass tore into their lungs and destroyed their respiratory systems. The rhinoceroses were the last to expire; their carcasses crowded the outer edge of the water hole. Interestingly, the same ash that created this scene of mass annihilation entombed it, forming a Pompeii-like time capsule of prehistoric life that is on display from May to mid-October in the Hubbard Rhino Barn of Ashfall Fossil Beds State Historical Park.

Since the site's discovery on the edge of an Antelope County cornfield in 1971, paleontologists have uncovered the remains of at least thirteen different mammals. There are numerous tooth-scarred, disarticulated bones, the result of predatory scavengers having savaged and scattered the carcasses. Many of the more than 350 skeletons that have been exhumed are in three-dimensional in situ posture. Llama-like camels, three-toed and single-hoofed horses, horned deer, four-tusked elephants, and saber-toothed cats exist in scenes of mass congestion. The remains of smaller animals, such as tortoises, birds, moles, and even an articulated snake, complete the catalogue of death.

One of the more surprising finds comprises bones of *Teleoceras*, a hippopotamus-like, stubby-legged, barrel-chested, herviborous rhinoceros that supposedly called both water and land home. Its remains form herdlike assemblages, many in moment-of-death poses. The five-to-one ratio of female to male skeletons suggests they lived in harem-like association. Two discoveries are perhaps unprecedented in the annals of paleontology: a female skeleton encasing the bones of her unborn offspring and a huddled mother and infant in a position suggestive of nursing.

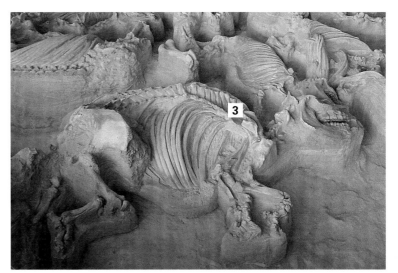

"Sandy" (center, #3) and her one-month-old calf "Justin" (upper right) were entombed with their noses touching. —Courtesy of Ashfall Fossil Beds State Historical Park

Lying side by side, partially exhumed rhinoceroses remain encased in the white volcanic ash that caused their death. The skulls of older individuals exhibit uneven tooth wear, giving them a snaggletooth appearance. —Courtesy of Ashfall Fossil Beds State Historical Park

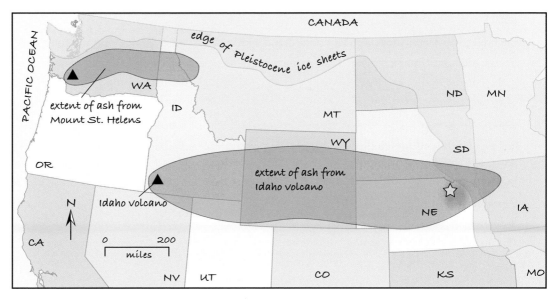

The Idaho eruption 11.83 million years ago spewed over seven states, an area several times larger than that affected by the 1980 eruption of Mount St. Helens. The ash beds at Ashfall Fossil Beds State Historical Park (star) exist today because they lay beyond the edge of the destructive ice sheets of the Pleistocene Epoch. —Modified after Rose et al. 2003

53. Little Blue River Outcrop, Nebraska

40° 10′ 21″ North, 97° 26′ 41″ West

Ancestors of present-day clams, oysters, and octopuses in a seaway depository.

In the multibillion-year history of Earth, four events stand out for their significance and intensity: the origination of life 3.5 billion years ago; the dramatic expansion of invertebrate species that lifted the curtain on the Cambrian Period; the Permian extinction episode that dropped the curtain on the Paleozoic Era; and the Cretaceous extinction event that felled as much as 85 percent of marine life and an equal percentage of all land species. Of these four life-changing incidents the last—occurring 66 million years ago—has the greatest notoriety.

As the Cretaceous Period came to an end, North America was teetering on the cusp of Armageddon. An asteroid as big across as Mount Everest is high and traveling twenty times as fast as a rifle bullet struck the Yucatán Peninsula of Mexico. Upon impact it released a billion times the energy of the atomic bombs that fell on Hiroshima and Nagasaki in 1945. The resultant atmosphere-fouling cloud of vaporized rock and dust blocked life-sustaining sunlight, delivered a deathblow to many forms of vegetation, and devastated the base of the food chain. Herbivorous dinosaurs starved, and with their demise their meat-eating brethren faced extinction.

North America at the time was geographically divided into two landmasses—Laramidia in the west and Appalachia to the east—by the Western Interior Seaway. Formed when a transgressing arm of the Arctic Ocean coalesced with an invading tongue of the ancestral Gulf of Mexico, the 600-mile-wide seaway extended from the flanks of the youthful Rocky Mountains to the foothills of the aging Appalachians. It is perhaps the greatest inland sea to have invaded North America. Tropical, relatively shallow, well oxygenated, and loaded with calcareous algae, the seaway was a warm-water larder that functioned as an incubator for evolving lifeforms, such as plesiosaurs, mosasaurs, and an amazing variety of mollusks—especially oysters and ammonites.

Less than 2 miles from Gilead, west along US 136 and then north for 1.7 miles on Road 6900 to the north side of the Little Blue River, the "little blue" outcrop offers ready access to the fossil remains of 90-million-year-old mollusks that went extinct after the meteorite impact. *Inoceramus* and *Exogyra*, two genera of oysters, are common in this sequence of gray Graneros Shale and overlying yellow-toned Greenhorn Limestone. Dubbed the "pearl oyster" of the Cretaceous because remnants of fossil pearls are occasionally discovered, *Inoceramus* has a thick shell identified by raised, semicircular growth lines. The largest bivalve in the fossil record, this oyster adapted to life in turbid water by developing a gill area large enough to allow it to cope with an oxygen-deficient environment. *Exogyra*, also thick and marked by a distinct configuration of ribs and pits that represent growth lines, is often referenced as the "devil's toenail" because of its gnarled, twisted shape. Both are commonly found in fragmented form.

Fossils of ammonites, the ancestral relative to the modern-day octopus and squid, are usually found as internal and external molds; their diameter ranges from that of a dime to that of a truck tire. Acclimated to the shallow waters of the Western Interior Seaway, they grew in tight spirals resembling fire hoses rolled for storage.

The fossils and sedimentary rocks of the "little blue" outcrop tell the tale of the geography-altering Western Interior Seaway, in existence for almost 40 million years. This vast body of water drowned central North America just before a rocky space wanderer forever changed the nature of life on Earth.

Gray beds of Graneros Shale (left center) overlain by yellow Greenhorn Limestone, both of Late Cretaceous age, make up the "little blue" outcrop.

The ribbed and spiral shell imprint of the extinct ammonite Eucalycoceras. *These predator cephalopods attacked with jet-propulsion movement and used a cluster of tentacles to grasp and consume their victims. Penny for scale.*

Some Inoceramus *species grew to a diameter of more than 3 feet. Fossils of these bivalves are ubiquitous within the Greenhorn Limestone, most commonly as fragments. Penny for scale.*

54. Berlin-Ichthyosaur State Park, Nevada

38° 52′ 26″ North, 117° 35′ 28″ West

A graveyard of undisturbed "fish lizards" entombed in 225-million-year-old rock.

After the dust of disorder and confusion of the Permian Period extinction had settled, the survivors desperately tried to return to once-familiar routines. It was not easy, for the biologic world had been transformed. The event's causes remain debatable—climate fluctuation, volcanic eruption, or bolide impact—but the long-term effects have long been known. It is the most life-defining catastrophe in Earth history: 96 percent of marine species became extinct, along with 70 percent of their terrestrial brethren. The myriad of animals that defined the Paleozoic Era was gone, replaced by a cavalcade of newcomers who would evolve and eventually dominate.

With the assembly of the supercontinent Pangaea completed, those species that successfully navigated the treacherous extinction event separating the Permian and Triassic Periods faced a future filled with opportunity. Vacated ecological niches abounded. On land, crocodiles, snakes, frogs, turtles, and lizards comingled with primitive mammals and dinosaurs. Offshore, emerging assemblages of snails, crabs, lobsters, and fish blanketed the ocean floor.

Pangaea's climate varied from moist and temperate in the polar and coastal regions to warm and arid in the interior. In contrast to the elevated basin-and-range topography that prevails today, Nevada was characterized by lowland environments: a shallow extension of the Panthalassa Ocean, the proto-Pacific, inundated the west, while scattered woods, marshlands, and lakes defined the plains to the east. Whale-like ichthyosaurs, humongous predatory reptiles, prowled the nearshore waters, driven by a never-ending hunger for squid and fish. Whether washed ashore by the gales of an unusual storm or grounded by an earthquake-induced tsunami, the carcasses of a pod of ichthyosaurs were flushed out of the water to rot in the sun. Today, this scene—preserved exactly as it was 225 million years ago—is the central attraction of Berlin-Ichthyosaur State Park, lauded as the largest and best ichthyosaur site in the world. The park, along with the nearby weathered remains of the late-nineteenth-century gold-mining town of Berlin, is located near Nevada 844 about 20 miles east of the community of Gabbs, in northwestern Nye County.

Paleontologists consider ichthyosaurs to be the most highly specialized reptiles that ever lived. From Early Triassic time to their extinction 90 million years ago, this alpha predator cruised the world's oceans with impunity. Related to a class of unidentified land cousins that returned to the sea, they had eyes the size of dinner plates—the largest of any known vertebrate—a feature that enabled them to hunt for prey in deep water. A cluster of thin, overlapping bones protected each eyeball from the pressures encountered at great depths. In spite of weighing more than 40 tons and being up to 70 feet long, they could move stealthily in excess of 25 miles per hour, propelling their smooth-skinned, streamlined body with limbs that had evolved into paddle-like flippers. They fed on a wide range of prey, as evidenced by fossilized stomach contents and analyses of fossil feces.

Ichthyosaur fossils have been found on every continent except Antarctica. The deposits at Berlin-Ichthyosaur State Park, a true "fish lizard" graveyard, contain the partially exhumed skeletal remains of at least nine animals, protected within the Fossil House. Skull, jaw, vertebra, flipper, and rib fragments of the genus *Shonisaurus*—adopted in 1977 as Nevada's official state fossil—lie in the exact orientation and position they occupied at the moment of death.

Cross-sectional exposure of the thorax of an ichthyosaur. Q is a short span of vertebrae that blends to the right into a bundle of ribs (R). The longest ribs are approximately 9 feet long.

A scattering of 6-to-8-inch-diameter vertebrae. Note that the front and back of each is hollowed out, a characteristic that differentiates ichthyosaur vertebrae from those of other marine reptiles.

Close-up of a seven-segment vertebrae column that composed a portion of a Shonisaurus *tail. Vertebrae are the most common fossils found at the visitors' quarry. Quarter for scale.*

55. Virgin Valley, Nevada

41° 50′ 02″ North, 119° 04′ 39″ West

Iridescent trace fossils that rival any opals found in the world.

Situated halfway between Adelaide and Alice Springs in the harsh desert of South Australia, Coober Pedy is perhaps Australia's most quirky town. Because a summer afternoon temperature of 117 degrees Fahrenheit is not at all unusual, most of the four thousand residents live and conduct their daily activities in *dugouts*, catacomb-like homes staked out in the miles of constant-temperature tunnels that underlie the treeless countryside. Three churches, a cluster of shops, a bar, and at least one private swimming pool provide evidence that this underground, sandstone-bound development has earned its reputation as a stable contemporary community, but in the first decade of the twentieth century the region was simply uninhabited country: forlorn, scorching, and enervating. Then, in 1915 significant deposits of opal—a hydrated form of silica—were discovered, men and machinery arrived, and steps were taken to transform the region into the opal capital of the world. The mines of Coober Pedy are world famous, but American gemstone enthusiasts can find high-quality opals much closer to home than the outback of Australia.

In contrast to its present-day sagebrush-blanketed, basin-and-range landscape, northwestern Nevada 16 million years ago was characterized by rolling hills, lakes, and lush forests of chestnut, hemlock, and spruce. Spasms of tectonic activity wracked the region, and the forests became smothered by multiple layers of volcanic ash. Over time, molecules of silica, deposited by migrating waves of groundwater, replaced the form and details of fragments of the buried wood. Uplift and erosion then altered the lay of the land, and the Virgin Valley, in Humboldt County, as well as its cache of gem-quality precious opal, came into being.

During the late 1800s cowboys, sheepherders, and prospectors traveling through the region occasionally stumbled across an exposed glittering opal. The first publicized discovery was in 1905—a full ten years before the Australian discovery—and soon afterward the Bonanza Opal Mine began production. Organized today as a privately owned "hobby mine" famous for opals radiating a multihued rainbow of color, the Bonanza is open seasonally on a fee basis. Most of the opal specimens are casts that filled molds left by decayed wood limbs, wood stems, bark, pinecones (rarely), and other types of plant material. These gem-quality iridescent casts can be classed as *trace fossils*—the indirect evidence of once-existing life, such as trails, tubes, and tracks.

Opals occur in a rainbow of colors—green, yellow, red, and blue—but the most fabled, and most treasured form of all, is the rare black opal, reported to be present in only two locations in the world: Australia and the Virgin Valley. The presence of carbon causes the black color, and iron oxide and other fine-grained trace elements create a vibrant play of colors, usually red or green. Combined, the colors and black background of these opals create a brilliance that far outshines the look of common opal. Quality specimens can be worth more than diamonds on a per carat basis.

To reach the Bonanza Opal Mine, from Denio Junction drive west on Nevada 140 for 21 miles, then travel south for 7 miles along the graded gravel road, following the signs to the mine site. Spectacular in appearance and rare, the highly collectible opal specimens of Virgin Valley should be front and center on any gemstone and fossil enthusiast's bucket list.

As can be seen by comparing a 16-million-year-old specimen (upper left) and a modern-day example (lower right), very little has changed with acorns. Penny for scale. —Specimens from the collection of Bonanza Opal Mines, Inc.

A 14-inch-long slab of opalized wood bark collected in the late 1940s from the Bonanza Opal Mine area of Virgin Valley. The iridescent display of colors is an effect typically associated with gem-quality precious opal. —Courtesy of Radford University Museum of Earth Sciences

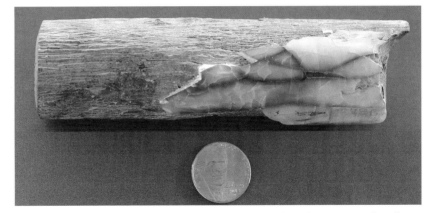

Limb-sized example of common opal, complete with preserved wood texture. Coin for scale. —Specimen from the collection of Bonanza Opal Mines, Inc.

A 4-inch-long, 10-ounce specimen of rare Virgin Valley black opal (left) composed of a dark-brown core and a nearly colorless outer layer. The back side (right) shows the original wood texture. —Specimens from the collection of Bonanza Opal Mines, Inc.

56. Odiorne Point State Park, New Hampshire

43° 02′ 30″ North, 70° 42′ 55″ West

A drowned forest in a state practically devoid of fossils.

One's degrees of success in finding fossils can be reminiscent of a country song: some days are diamonds, others more like stones. A diamond day is the discovery of a *lagerstätte*, a fossil assemblage noted for its remarkable diversity and quality of preservation. American examples include the biota of Mazon Creek, Illinois (site 27), and the fish of the Green River Formation in Wyoming (site 99). A day of stone is when nothing is found, for a variety of reasons. Nobody expects to find fossils in volcanic terrain, as the high temperature of its formation excludes the presence of life, and many a sedimentary rock is barren because it was deposited in an environment either not conducive to life or lacking in effective conditions for fossilization. Finally, there are areas that were a "diamond," but in the geologic past the rock-crushing and annihilating effects of metamorphism and igneous intrusion turned them to "stone." New Hampshire is a prime example.

The paleontologic story of the Granite State can be summarized in a few short paragraphs using phrases such as "fossils are very rare," "discovery is a difficult task," and "their presence is at best cryptic." For the better part of 300 million years—throughout the Paleozoic Era—the tumultuous forces associated with the convergence of the protocontinents of Africa and North America, the closing of the proto–Atlantic Ocean, and the development of the Appalachian Mountains impacted New Hampshire. By the time this prolonged episode of change was over, sandstone, shale, and limestone formations had been altered to quartzite, slate, and marble, respectively; the crust had been compressed horizontally as much as 600 miles; and the countryside was peppered with massive intrusions of molten rock. Any fossils that survived this history of construction and destruction—if they exist at all—have yet to be discovered.

However, by fast-forwarding to the later stages of the Quaternary Period, there is a fossil story to be told at Odiorne Point State Park. During the height of the Pleistocene Epoch 14,000 years ago, the weight of thousands of feet of glacial ice depressed the land. As temperatures warmed, however, melting of the ice caused land to rise locally and sea levels to rise worldwide. Initially sea-level rise outpaced the rebounding process, but eventually the land emerged relatively high and dry and became blanketed by white pine, hemlock, and maple firmly rooted in peat-rich woodland soils 2.5 to 4 feet thick. However, sea-level rise outpaced land rebound again and slowly drowned the forest. Today its remnants can be observed at low tide in the 650-foot-diameter tidal pool cove immediately south of Odiorne Point.

The fossil forest, only partially intact, is composed of abraded stumps, fallen logs, and the whorled remains of root systems preserved in a fibrous condition by the brine of the Atlantic Ocean. These are not the petrified specimens so common to many other fossil-wood localities. Radiocarbon age dating indicates the trees were drowned during a 1,300-year period, from 4,500 years ago to 3,200 years ago. Analyses of samples taken from the outer rings of four white pine stumps show that regional sea levels were rising during that time period at an average 0.31 feet per century. This scientific evidence, deduced from the fossil forest of Odiorne Point State Park, indicates that sea level, whether exacerbated by the lifestyles of humankind or not, has been rising for a very long time.

At low tide the fossil tree stumps of Odiorne Point are partly exposed and partly covered by residual pools of water and beach pebbles. The root system of this specimen is exposed to the forces of tidal erosion. Dollar bill for scale. —Courtesy of Jim Cerny

Despite being thousands of years old, Odiorne Point fossil wood is so beautifully preserved that tree ring analysis is often possible. —Courtesy of Jim Cerny

57. Big Brook Preserve, New Jersey

40° 19′ 11″ North, 74° 12′ 51″ West.

A nationally known site for collecting shark teeth and excrement.

Every year some 200 million men, women, and children visit the beaches of the United States. Many wish to soak up some rays, while others arrive loaded with devices designed for activity: scuba masks, wet suits, snorkels, and surfboards. Few consider the possible risks associated with a day at the beach, and fatalities caused by surfing and other beach-related activities are normally accepted as yesterday's news. Reports of a shark attack, however, whether fatal or not, are cause for alarm.

Sharks have gained an increasingly less-than-desirable reputation, perhaps related to the 1975 movie *Jaws* and endless shark-attack documentaries on television. About seventy-five attacks are reported worldwide annually, but less than 6 percent prove fatal, indicating that the bum rap they are given is misdirected. Nevertheless, many people fear sharks more than any other fish, and shark teeth are prized specimens in many fossil collections.

Agnathids, the very first fish, date back to the Ordovician Period. These primitive and jawless creatures, related to today's lamprey eels, gave rise to fish with jaws and then, during Devonian time, to fish with an internal framework composed almost entirely of cartilage. With this development the Age of Sharks was born.

Shark cartilage—tough, white, flexible, and fibrous tissue like the material human ears and noses are made of—is much softer than bone and does not fossilize well. It decomposes quickly after death, leaving a residue composed of teeth and scales. The root and core of the teeth are composed of true bone covered by enamel, a hard calcareous substance that makes the teeth resistant to fracture and abrasion.

Shark teeth are designed to be replaced when they are lost or broken by normal activity, such as attacking prey or eating a meal. Teeth are arranged in rows, the number varying from species to species. Whenever a tooth is lost, the tooth in the row behind it moves forward, generally within a week's time, to take the missing tooth's place. Since any number of teeth can be lost and replaced, any one animal may use ten thousand to fifty thousand teeth in a lifetime. This fact explains why shark teeth are so very common in many sedimentary formations.

Big Brook Preserve is famous as a supermarket of shark teeth, where at least ten varieties can be found. From Montrose proceed 0.8 mile north along Boundary Road to Vanderburg and Crine Roads, then east 0.5 mile to Hillsdale Road, and finally north 0.6 mile to the parking lot (95 Hillsdale Road). In addition to teeth, with patience, a sharp eye, and dedication to the hunt, the fossil aficionado can find fossilized shark excrement, or coprolites. Ranging up to 2 inches in length, these trace fossils are easily recognized by the fact that no two specimens have the same surface ornamentation or overall shape and size. Extinct belemnites, a variety of vertebrate bones, and the fossil remains of Cretaceous-age reptiles can also be found at Big Brook, but identifying them can be difficult due to their fragmented nature and similar appearance.

Big Brook is a most unusual trove of fossils, a place where the adventuresome visitor has the rare opportunity to collect 75-million-year-old specimens from a shark's two working ends. Here you can find teeth and excrement of these ancient predators of the depths, aptly described as nature's most perfect killing machines.

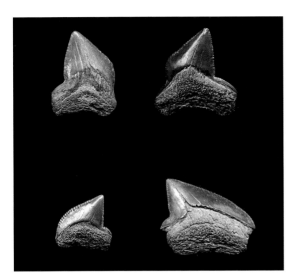

Squalicorax, *the extinct 15-foot-long crow shark, was a coastal predator, but it also scavenged, as evidenced by a "squali" tooth found embedded in a dinosaur foot bone. This is a very common tooth at Big Brook. The small tooth at lower left is 0.5 inch wide.* —Courtesy of Jayson Kowinsky

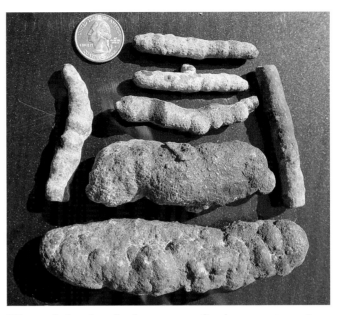

Wherever shark teeth are found, so too are coprolites; however, no two specimens are the same size or shape. Quarter for scale. —Courtesy of Jim Lacefield

A sampling of the variety of shark teeth and bits and pieces of vertebrate bone found at Big Brook. Most specimens average 0.5 inch in size and are best collected with a sifting pan that has a 0.25-inch mesh screen. —Courtesy of Jayson Kowinsky

58. Hamburg Stromatolites, New Jersey

41° 08′ 35″ North, 74° 34′ 40″ West

Ancestral bacteria colonies born in an energized realm and molded by glaciers.

Many of the problems associated with the study of geology would be unsolvable without paleontology. The presence of fossils in rocks is evidence of a variety of preexisting conditions, such as depositional environment, biological habitat, and degree of tectonic activity. The geologic range of any fossil—its life span, with circumstance of evolution and extinction serving as bookends—is of special value. The trilobite, a type of arthropod that thrived in the Paleozoic Era, demonstrates the usefulness of this range. Any trilobite-bearing sedimentary rock found anywhere in the world is known to have been deposited sometime between 541 and 252 million years ago. If the collective range of the worldwide catalogue of fossils extended from the dawn of the Archean Eon to the sunrise of tomorrow, their value as evaluators of age would be reduced to zero.

The geologic range of any animal or plant fossil is considered valid until older or younger specimens are discovered. The history of stromatolites is a good example. These fossils formed in aquatic environments as biofilms of cyanobacteria accreted layered mats of fine-grained sediments. Once considered extinct, living examples of these curious structures were found flourishing in the salty water of Shark Bay, on the extreme west coast of Australia, in 1956. Further range adjustment occurred with the discovery of inferred stromatolites in rocks estimated to be 3.465 billion years old. This find gave stromatolites the title of the world's oldest fossils.

Unusual and illustrative examples of stromatolites occur in an outcrop at the southern end of the loop road through Ballyowen Estates, 0.3 mile west of the intersection of Gingerbread Castle Road and New Jersey 23, immediately south of Hamburg. The rock is a hump of middle-Cambrian-to-Early-Ordovician-age Allentown Dolomite with a polished surface that invites up-close-and-personal inspection. Clusters of spherical stromatolites, ranging in diameter from small to more than 3 feet, are intimately associated with desiccation cracks and oolites, both features indicative of a shallow water depositional environment.

Oolites are constructed of concentric layers of calcium carbonate that accrete around a nucleus, such as a shell fragment or an algal pellet. They are precipitated in tropical waters and are defined as being smaller than 0.08 inch in size. Their spherical to oval shape is the result of to-and-fro agitation in wave- and storm-energized water. Desiccation cracks form when either mud or carbonate ooze is exposed to the atmosphere, causing the sediment to dry out and crack. These features, along with the acknowledged geologic age, are evidence the Hamburg structures formed some 500 million years ago in warm and shallow waters. This interpretation is supported by plate tectonic history. At the time Laurasia, the embryonic North American continent, was situated between 15 and 20 degrees south latitude—a tropical locale, indeed.

During the Pleistocene Epoch, glacial erosion sculpted the Hamburg exposure into a *roche moutonnée*—French for "fleecy rock"—an extended knob of bedrock elongated in the direction of ice movement. The orientation of the many glacial striations, or scratches, that tattoo the smooth surface is the same as that of the long axis of the knob, definitive evidence that ice muscled its way across the countryside in a southwesterly heading.

Born 0.5 billion years ago in an energized and warm environment and then reconfigured in the frigid clime of the ice age, the photograph-only Hamburg stromatolites may be the finest specimens of their type in the United States.

Hamburg stromatolites are principally spheroidal in shape. Studies show that three-dimensional organic structures of this type have a tendency to live and grow in high-energy aquatic environments. Striations (white streaks trending left to right) are convincing evidence that glacial ice sculpted this exposure. Pen for scale.

Desiccation cracks in the carbonate rock separating two Hamburg stromatolite mounds (upper left) offer evidence that these cyanobacterial structures lived within a shallow-water intertidal environment.

From a distance, the smoothed and rounded, whaleback-shaped roche moutonnée configuration of the Hamburg dolomite exposure is obvious.

59. Battleship Rock, New Mexico

35° 49′ 46″ North, 106° 38′ 46″ West

A nautical-named promontory that is home to oceanic fossils.

The geologic story of New Mexico, a tale that extends from the archaic years of Precambrian time to volcanic episodes of the Holocene Epoch, has many diverse chapters. However, the chapter on Battleship Rock, a volcanic promontory defining the skyline of the Jemez Mountains in Sandoval County, has long garnered particular attention. This locale is lauded for containing the most fossiliferous strata in the entire state.

Early on, starting circa 1.6 billion years ago, the continental landmass today known as the Land of Enchantment could best be described as the "land of upheaval and distortion." For millions of years ancestral highlands eroded, resulting in immense thicknesses of sediment that were compressed into a sedimentary foundation, which was then metamorphosed and intruded by magma. For the ensuing 1.3 billion years, the history of this Precambrian core was shrouded in the mists of time. The 1.3-billion-year episode of missing history, known as the Great Unconformity—the most famous time gap in the geologic record—separates rock containing familiar varieties of fossils from rocks generally void of fossils. This unconformity can easily be traced throughout the Rocky Mountains, from the depths of the Grand Canyon to north of Wyoming.

Then, 300 million years ago, in the Pennsylvanian Period, the land was inundated by a shallow ocean teeming with bryozoans, crinoids, mollusks, and brachiopods. Tectonically, the Pennsylvanian Period was highlighted by the final assembly of globally distributed landmasses into the colossus Pangaea, but it was also a time of minimal relief and universal high sea levels. Sediment that eroded off the flanks of the rising ancestral Rocky Mountains was deposited in the shallow waters covering the Jemez Mountain area and compressed into fossil-rich strata of the Madera Group. Overlying this alternating multitude of layers of marine limestone and shale are river-deposited sandstone and siltstone of the Abo Formation. As a whole, this sequence of marine rock blending upward into terrestrial rock is prime evidence that the oceanic shoreline zigzagged across the Jemez Mountains area for a considerable amount of time before the rising Rocky Mountains pushed the sea south.

During the Pleistocene Epoch, sporadic episodes of volcanic activity wracked central New Mexico. Blankets of ash and rock spread across the land, and heat and pressure welded them into ignimbrite, an igneous rock formed by the explosive discharge of a volcano. Subsequent erosion of this uncommon rock formed a variety of landforms, the most interesting of which is Battleship Rock, a promontory that fills a vertical gorge sculpted into strata of the Madera Group and the Abo Formation.

Most fossils associated with the Madera Group have weathered out of the Jemez Springs Shale Member, a steep-sloped, high-off-the-road exposure that is generally accessible to only the most physically fit of weekend paleontologists. The more prudent visitor can examine the rubble found along several streambeds that punctuate the prime collecting area, which extends along New Mexico 4 several miles north and south of the Battleship Rock parking area, about 33 miles west of Los Alamos. Crinoid fragments are common and gastropods and bryozoans can be found, but the prized remains are any genera of brachiopod, especially those of *Punctospirifer*, *Hystriculina*, the hard-to-find *Neospirifer*, and the common to abundant *Composita*.

Composita, *a genus of brachiopod that lived from the Carboniferous to the Permian Period, is relatively abundant in Madera Group strata at Battleship Rock. The shell is small and smooth and distinguished by a circular opening that once housed a muscular appendage that anchored the animal to the seafloor. Dime for scale.*

Enveloped by the haze of a developing forest fire (center right), the volcanic Battleship Rock has been permanently anchored within fossil-bearing sedimentary rock since it formed 60,000 years ago.

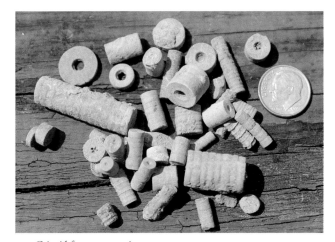

Crinoid fragments consist of elongated stem segments and individual wheel-shaped columnals that are created when stems break apart. Dime for scale.

The richest fossil-bearing zones are located within difficult-to-reach shale beds forming steep slopes along the west side of New Mexico 4 in the vicinity of Battleship Rock.

60. Ghost Ranch, New Mexico

36° 19′ 48″ North, 106° 28′ 26″ West

Triassic-age dinosaur and reptile fossils in the gorgeous Red Rock Country.

Spread across 130,000 square miles of Colorado, Utah, Arizona, and New Mexico, the Colorado Plateau, widely known as Red Rock Country, has been a relatively stable chunk of Earth's crust for the better part of 600 million years. Plateau terrain is characterized by a thick, breathtaking array of pink, yellow, brown, orange, red and white sandstone, shale, and limestone that is minimally deformed. Throughout the Paleozoic Era the region was periodically inundated by tropical seas, but then, 250 million years ago, it assumed a terrestrial nature created by tectonic uplift related to the amalgamation of multiple landmasses into the supercontinent Pangaea. The fossil catalogue reflects this transformation, changing from one composed principally of marine invertebrates to a collage of air-breathing vertebrates.

Red Rock Country is the terrain from which Georgia O'Keeffe, the mother of American modernism, collected bones and rocks that she incorporated into famous depictions of the New Mexico landscape. *Ram's Head, White Hollyhock, and Little Hills*, an iconic 1935 painting, epitomizes her love of Red Rock County, extant flora, and expired fauna. In 1940 she moved into a house about 14 miles northwest of the small community of Abiquiú, on land occupied by the Ghost Ranch, today an educational and retreat center that includes several quarry sites that preserve the remains of an interesting array of Triassic-age vertebrates.

Rock exposures in the vicinity of Ghost Ranch range from brick-red and white to tan siltstones and sandstones of the Triassic-age Chinle Formation to the yellow tones of the younger Jurassic-age Entrada Sandstone. Renowned fossil discoveries made in local quarries include the skeletal remains of *Coelophysis*, an early dinosaur; a skull of *Vancleavea*, an eel-like reptile; *Tawa*, a carnivorous dinosaur; and *Effigia okeeffeae*, a reptile named after Georgia O'Keeffe.

Almost completely enveloped with at least four different types of armor plating covering different parts of its body, *Vancleavea* is believed to have shared its aquatic environment with larger and more powerful neighbors, thus the need for protection. Like crocodiles and phytosaurs, this 4-foot-long, extinct reptile had eyes and nostrils on top of its skull, allowing it to see and breathe while the rest of its body was submerged. This unusual egg-laying vertebrate had an elongated body, thick bones, small limbs, and a dense and robust skull.

Fleshed out around a skeleton constructed of hollow bones and a true wishbone, *Coelophysis* is not only a distant ancestor of modern birds but also one of the best-represented meat-eating theropods in the fossil record. About 8 feet long and weighing 40 pounds, it ranged the Colorado Plateau long before the golden age of the dinosaurs. Aided by large, forward-facing eyes that supposedly afforded it excellent depth perception, this predatory dinosaur had little difficulty satisfying its daily intake of calories. Thousands of *Coelophysis* bones have been discovered at Ghost Ranch, perhaps the remains of individuals killed during a flash flood.

Locally discovered and researched fossils are on display at the Ghost Ranch Museum of Paleontology, along with a collection of instruments used by fossil enthusiasts in the field when extracting specimens. From delicate dental picks to a foot-long hammer, the tool kit includes a small ruler that some deem necessary to keep paleontologists honest. After all, a few individuals are prone to telling fish stories while describing the size of their fossil discoveries.

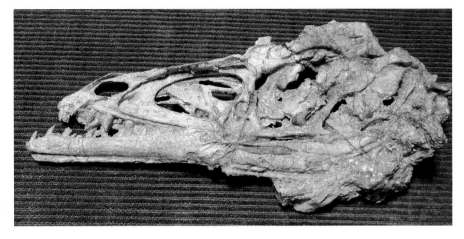

Sharp as a well-honed blade and jagged with minute serrations on both edges, the teeth of Coelophysis *were well suited for this dinosaur's carnivorous and predatory lifestyle.*
—Courtesy of Alex Downs, Ghost Ranch Museum of Paleontology

Vancleavea's *skull is characterized by nostrils that face upward, suggesting it could breathe by sticking its head above water like a crocodile, and large, powerful teeth designed to penetrate tough prey, such as heavily scaled fish.*
—Courtesy of Alex Downs, Ghost Ranch Museum of Paleontology

Located along US 84, 2 miles south of the entrance to Ghost Ranch, this imposing exposure includes strata from both the Chinle Formation (lower three-quarters) and the younger Entrada Sandstone (upper massive unit). Coelophysis *bones are found in the youngest beds of the Chinle Formation.*

61. Clayton Lake State Park, New Mexico

36° 34' 40" North, 103° 17' 47" West

Hundreds of imprints record a once busy iguanodon freeway.

When an earthen dam was built across Seneca Creek in the 1950s, creating Clayton Lake, north of Clayton, no one had any idea that its crown jewels lay so close to the surface. In 1982, however, floodwaters eroded a layer of silt from the spillway, revealing a bonanza of dinosaur tracks for the first time in 100 million years. Today this collage of five hundred ichnofossils is one of the best-preserved and most extensive dinosaur track sites in the United States.

The tracks were left along the western shoreline of the Western Interior Seaway, a shallow sea that formed during Cretaceous time when rising waters of the Gulf of Mexico joined those of the Arctic Ocean. Sediment deposited in the ocean, as well as in the rivers flowing into it, was the ideal canvas for the dinosaurs; it later hardened into sandstone and shale. While a plethora of dinosaur tracks have been found from northern Colorado to central New Mexico, no dinosaur bones have yet been found. The tracks are the sole evidence that dinosaurs ever inhabited the area.

Ninety percent of the Clayton Lake tracks consist of broad three-toed impressions ranging in size from that of the shallow, 1-inch-long imprints of a juvenile to the deep, 1-foot-long indentations of adults. Paleontologists are able to determine the category of dinosaurs that made these tracks, but they are unable to relegate them to a specific genus, thus they have been given the scientific designation *Caririchnium*. Most of the tracks were left by ornithopods, 3-foot-long, bough and shrub grazers that, over time, grew in size and numbers until they became the most successful group of herbivores of the Cretaceous Period. The remaining 10 percent of the tracks, characterized by pointed toes and a thinner footprint, were primarily formed by theropods, bipedal meat-eating dinosaurs.

Experts believe the most common Clayton Lake tracks were made by iguanodons, robust ornithopods with a large, narrow skull and teeth that resembled those of iguanas. These bulky creatures were distinguished by powerful hind legs and feet constructed of three relatively long toes. Track analyses suggest the iguanodon was primarily bipedal but could, if required, easily move about on all fours. When on two feet, it was capable of running 15 miles per hour.

Individual Clayton Lake tracks, the longest of which is twenty steps in length, are sited along a single line, indicating the animals walked with their legs planted beneath their bodies and not extended out laterally like modern reptiles. Examples of parallel trackways are also present, evidence that some dinosaur species, and perhaps varying age groups of the same species, traveled together along a defined migration route.

Iguanodon tracks are found as two distinctive shapes, the result of the Clayton Lake mudflat having different water content at different times. When the mud was viscous, sharp, well-defined tracks were left in the mud, clearly outlining the three toes of the dinosaur's hind foot. When it was less viscous, however, a triangular kite-shaped imprint formed, caused by mud settling back into the imprint after the animal had moved forward. All the tracks were left in shallow water, as evidenced by the presence of worm burrows, ripple marks, and mud cracks. Multiple boardwalks and a gazebo housing a variety of informative plaques allow for a leisurely and educational examination of the Clayton Lake trackway. Located 12 miles north of Clayton, this iguanodon freeway is a must-visit site for dino enthusiasts of all ages.

Located to the right of the gazebo, these generally hexagonal 100-million-year-old mud cracks provide indisputable evidence that northeastern New Mexico was a shallow, nearshore aquatic environment during the Cretaceous Period. The largest mud crack in photo is 10 inches in diameter.

Well-defined, 1 inch in depth, and preserved for 100 million years, this imprint—supposedly left by an iguanodon—is a common footprint at Clayton Lake State Park. Pen for scale.

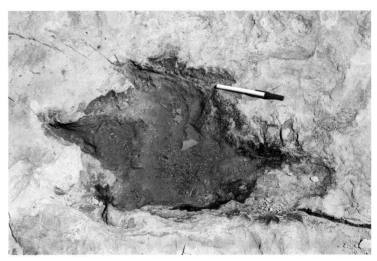

The length of each toe and the suggestion of claw marks suggest this footprint was left by a meat-eating theropod. These are the minority dinosaur tracks at Clayton Lake. Pen for scale.

62. John Boyd Thacher State Park, New York

42° 39' 01" North, 74° 00' 22" West

Following the trail of pioneer geologists along a fossil-rich escarpment.

The second wave of immigration from the Old World to the New World began in 1830—the first had ended in 1775—and peaked two decades later when 1.7 million Europeans moved into the Midwest. Motivated by hopes of freedom and opportunity, most travelers sought to build a new life around family and land, but a few were energized by controversial geologic hypotheses focused on New York State: disputes dealing with paleontology, the role of glaciation, and the concept of uniformitarianism in the interpretation of Earth history. Armed with a copy of Williams McClure's 1809 map—the first readily available representation of the geology east of the Mississippi River—they traveled up the Hudson River to Albany and then inland to the Helderberg Escarpment.

Stretching for 10 miles and looming upward of 900 feet, the escarpment dominates a vista defined by the valleys of the Mohawk and Hudson Rivers, the southern Adirondacks, and the western ranges of Vermont and Massachusetts. The centerpiece of this colossal wall of rock is protected by the boundaries of John Boyd Thacher State Park, west of Albany. The trails tourists walk here were pioneered during the early decades of the nineteenth century by a litany of scientists, each a "founding father" of a specific discipline: Louis Agassiz (glaciology), James D. Dana (mineralogy), James Hall (paleontology), and Sir Charles Lyell, one of the world's most influential geologists. The writings of these trailblazers refined New York State geologic history to the point that it became the key to understanding the origins of North America.

One well-trodden pathway, the Indian Ladder Trail, dates back to a time when Native Americans supposedly scaled the final 30 feet of the escarpment using a notched white pine log. A walk along this 0.5-mile-long trail from east to west is a journey back in time to 418 million years ago, when the park lay 30 degrees south of the equator, awash with transparent, tropical waters. Two distinct photograph-only-and-take-no-samples formations make up the bulk of the escarpment: the thin-bedded Manlius Limestone and the overlying, younger, massive 36-foot-thick Coeymans Formation limestone.

Manlius Limestone is recognized by three distinctive fossils: *Tentaculites*, 0.5-inch-long, carrot-shaped, mollusk-like shells; *Howellella*, a type of brachiopod; and *Leperditia*, a small bean-shaped ostracod. The parallel alignment of *Tentaculites* is indicative of life and death in a sometimes turbulent tidal-flat environment; the rushing tides oriented the fossils in a direction paralleling their flow. Mud cracks, occasional ripple marks, and sporadic stromatoporoids and stromatolites provide further proof of nearshore deposition. *Stromatoporoids* are layered, cabbage-shaped, extinct masses related to present-day sponges, and *stromatolites* are sediment mounds built by colonies of cyanobacteria.

Exposures of the Coeymans Formation, best viewed at the west end of the trail along the sixty-one-step ladder, tell a different story. The most common fossils found here are thick-shelled *Gypidula* brachiopods and several *Favosites*, a honeycomb coral so named because of its resemblance to hexagonal, thin-walled honeycombs. This diversity of broken and abraded fossils is evidence the Coeymans was deposited in a high-energy, wave-dominated environment offshore a barrier island constructed by corals. This environment contrasts sharply with the nearshore conditions that prevailed when the Manlius was being deposited.

One of the richest fossil-bearing exposures in the eastern United States, the Helderberg Escarpment is rockbound evidence that the sea-level rise experienced today, attributed to global warming, is a reoccurring phenomenon. This fact, preserved in the escarpment's geologic history, serves as another persuasive example of what goes around comes around.

Mine Lot Falls along Indian Ladder Trail showing well-defined strata of the Manlius Limestone (bottom) overlain by massive layers of the Coeymans Formation. —Courtesy of Charles Ver Straeten, New York State Museum

(Inset) *A memorial to the many pioneering geologists who worked along the Helderberg Escarpment in the early nineteenth century. Note the decorative edging of fossil images.* —Courtesy of Charles Ver Straeten, New York State Museum

slab of Manlius Limestone full of Tentaculites. The distinctive alignment is reflective of the direction of tidal flow at the time these extinct mollusk-like fossils were deposited. Centimeter ruler for scale. —Courtesy of Charles Ver Straeten, New York State Museum

The thick shells of these Gypidula brachiopods protected the living animal from the high-energy environment in which the Coeymans Formation was deposited. Pen tip for scale. —Courtesy of Charles Ver Straeten, New York State Museum

63. Penn Dixie Fossil Park and Nature Reserve, New York

42° 46′ 38″ North, 78° 49′ 51″ West

Fossils large and small, including a few of "gold standard" quality.

The view from the edge of American Falls, on the American side of Niagara Falls, north of Buffalo, is acclaimed worldwide in terms of tumultuous force and irreversible destruction. Every second 75,000 gallons of water cascades over the precipitous incline, plummets at the rate of 32 feet per second, and attacks the underlying talus slope with a force of 750 tons. Since its birth 12,000 years ago its lip has eroded upstream at an average rate of 3 feet per year—on occasion 6 feet per annum—evidence it may indeed be, as claimed, the fastest retreating waterfall in the world. To the tourist, American Falls is unprecedented in terms of height and discharge; to the geologist it is a textbook example of physical brute power.

About 30 miles to the southeast, surrounded by woods to the north, east, and west, there is a landscape of a very contrasting nature. To the casual observer it appears gray, lifeless, and moonscape-like, but the practiced eye sees a venue of bygone fertility and evolutionary history. Located 0.75 mile west of I-90 at 4050 North Street in Blasdell, and open seasonally to the public on a you-find-it-you-keep-it basis, the central asset of the Penn Dixie Fossil Park and Nature Reserve is a 33-acre exposure of sedimentary strata rich in Middle Devonian fossils. An astonishing variety and volume of corals, brachiopods, crinoid columnals, pelecypods, gastropods, and cephalopods are embedded in gray shale of the Wanakah Formation and Windom Shale. The main attractions, however, are the well-preserved, thumb-sized trilobite specimens of the genera *Phacops* and *Greenops*, the former noted for a distinctive eye structure and the latter for spiny, body-encircling serrations.

Penn Dixie has been in the development stage for 390 million years, dating to the time when the Acadian orogeny, the second of the three phases of mountain building that were instrumental in the development of the Appalachian Mountains, was ending. New York State was then located 10 degrees south of the equator and flooded by warm, shallow waters of the Kaskaskia Sea, a dumping site for sediment eroded from the uplifted Acadian highlands to the east. Embayments lacking both oxygen and predator life became overcrowded graveyards, as evidenced by numerous trilobite exoskeletons preserved in a variety of death poses: some clustered and overlapping each other, some rolled into ball-like form, and others lying in a prone position. Penn Dixie is one such locale.

The pygmy fossils of the pyrite bed section of the Penn Dixie site merit special attention. To the novice these pea-sized trilobites, brachiopods, and cephalopods may appear to be sculpted from gold bullion, but they are really the final product of *pyritization*, the process by which the original fabric of a fossil is covered with pyrite, an iron-bearing mineral also known as fool's gold. Pyritized fossils are evidence that organisms died in an oxygen-deficient environment charged with a high concentration of dissolved iron. The involvement of sulfur-reducing bacteria is a key factor in the formation of the gold-like luster. Prized as "gold standard" fossils by many collectors, pyritized fossils are, unfortunately, susceptible to "pyrite disease," a process in which pyrite chemically decays to a brass-colored sludge when placed in a humid environment.

Considered one of the richest deposits of Devonian-age exoskeletons in the eastern United States, Penn Dixie Fossil Park is a prime example of timeless biologic preservation and a sharp contrast to the ever-present forces of geologic destruction underway 24/7 at nearby American Falls.

These 0.2-to-0.25-inch-diameter pygmy "gold standard" fossils were gradually altered to fool's gold in a bacteria-rich, oxygen-deficient environment when iron sulfide replaced the calcium carbonate shells of the deceased cephalopods. —Courtesy of Tasha Mumbrue

Distinguished by a globular head and large, bulging eyes, Phacops *is often found rolled up, a biologic defense tactic that is closely associated with this genus of trilobite. The prone form of this particular detritus eater, which could grow up to 6 inches long, is the version most treasured by fossil enthusiasts.* —Courtesy of Tasha Mumbrue

The fragmented remains of at least six Phacops *trilobites in this 8-inch-long slab are suggestive of death by natural disaster, such as burial by a turbidity current or entrapment in an oxygen-deficient environment.*

A partially exhumed Greenops *from the Windom Shale at Penn Dixie. Averaging 1 to 1.5 inches long, this genus of trilobite lived in a warm, fairly deepwater environment, commonly in association with* Phacops. —Courtesy of Tasha Mumbrue

64. Aurora Fossil Museum, North Carolina

35° 18′ 17″ North, 76° 47′ 15″ West

A museum and fossil pit focused on a top marine predator with giant teeth.

Shark teeth are found on beaches all along the southeastern coast of the United States. Some have an age measured by geologic time, but many are derived from hosts that are still trolling the depths of the oceans in search of food and a mate. Somewhere around 460 living species of sharks have been described, and more than 50 occupy estuaries, the continental shelf, and deep waters of the North Carolina coast. All share a common characteristic: they continually shed teeth.

Sharks replace their teeth on a periodic basis. For each tooth lost, as many as six exist in successive stages of development, ready to move into a front-row position in the predator's mouth. The total number employed in a lifetime by any one individual depends on the species. For example, the infamous white shark uses some thirty thousand teeth in its thirty years of existence. As such, one can begin to understand why fossil collectors often declare that shark teeth are the most common type of fossil to be found.

The most famous shark teeth, considered by many paleontologists to be the crown jewels of their collection, belong to *Carcharodon megalodon*, a species designation meaning "big tooth." Weighing in at more than 60 tons and measuring more than 65 feet in length, Megalodon, or "Meg," sharks occupied the alpha position of the oceanic food chain for a 20-million-year reign of terror during the Miocene and Pliocene Epochs, and then they became extinct. With an appetite as big as its reputation—that is, as the most successful carnivore that ever lived—Megalodon is believed to have fed primarily on whales, first biting off their fins and then plunging in for the kill while the victims struggled to escape. Secondary targets included dolphins, dugongs, and giant sea turtles, as evidenced by bones with bite marks that match Megalodon dentition.

Even novice paleontologists can easily recognize Meg teeth. Triangular in shape and up to 7 inches long, their robust structure is characterized by fine sawtooth serrations and a massive V-shaped crown. Larger than any other shark teeth, they were employed with a bite force that exceeded 12 tons per square inch—3 times greater than that of *T. rex*.

The assembly of any Meg tooth collection should begin at the Aurora Fossil Museum (400 Main Street) in Aurora. Visitors are encouraged to first explore for mako, snaggletooth, dusky, and bull shark teeth—as well as specimens of shark coprolites—in the "Pits of the Pungo," a landscape of Pungo River Formation spoil piles that contain multitudes of Miocene fossils. Located across the street from the museum, the pits are replenished twice a year with fossil-rich sediment and are open to diggers and prospective paleontologists of all ages.

Many visitors consider the "shark gallery" the must-see exhibit of the museum. This display of fossil memorabilia from the nearby Aurora phosphate mine is arguably one of the most important sources of Miocene and Pliocene fossils in the world. The far wall of the gallery—a world-class review of pristine, perfectly formed Meg teeth and shark dentition—contrasts with an adjacent exhibit of pathological shark teeth deformed by mutations, injuries, and infections. These two collections display stark evidence that Megalodon, the onetime savage brute of the marine world, had, like all creatures, its good and bad days, including those when its cavernous mouth functioned with razor-sharp efficiency and those when it suffered from an enervating and throbbing toothache.

Teeth specimens are what principally represent Carcharodon megalodon, *the extinct giant white shark, in the fossil record. The relationship of this humongous predator to the present-day great white shark remains a matter of considerable controversy. Tooth at the top center is 4 inches long.*

Shark coprolites—a technical term for fossilized feces—exist in a variety of sizes and shapes. These 5-to-20-million-year-old examples are typical of the specimens found in the Miocene Epoch formations of North Carolina. Quarter for scale.

Tusks are the most prominent feature of the walrus. A walrus probably employed this broken 8-inch-long tusk to form and maintain entry holes in ice and as an aid for climbing out of the water onto ice.

Best known for its world-class shark fossils, the Aurora Fossil Museum hosts a range of related marine exhibits. The apparent absence of teeth in the beak-like snout of this 8-inch-long dolphin skull is mystifying.

65. Green Mill Run, North Carolina

35° 36′ 22″ North, 77° 20′ 36″ West

In the film, fossil, and real worlds, the great white is the most macho of predators.

For more than two hours the suspense increased. Five innocent locals met gruesome deaths, city officials debated the what and when of retaliation, and millions of viewers were frightened out of their seats. When the final curtain descended on more than 450 screens nationwide and moviegoers' nerves had returned to a normal state, the 1975 summer blockbuster *Jaws* had attained the high-water mark in motion picture history as the highest-grossing film of its time, with box-office receipts totaling nearly $500 million. Before *Jaws* the American sun-seeking public was generally more concerned about finding an onshore place to spread a blanket than the presence of sharks offshore. After *Jaws* many a weekend beachgoer hesitated before dipping even a toe into the water.

Classified by the scientific community as *Carcharodon carcharias*, the great white is known to most ichthyologists as the world's largest predatory shark. Others think of it simply as 2 to 3 tons of muscle, munch, and meanness. The oldest known fossils date to the midpoint of the Miocene Epoch, 14 million years ago, but the great white's origin is a matter of debate. An initial hypothesis suggested it shared common ancestry with the extinct shark *Carcharodon megalodon* (see site 64), but more recent studies relating to size and teeth similarities indicate these marine marauders are best considered distant cousins.

The reputation of the great white is based on its prowess, habitat, and distribution. Preferring waters around 65 degrees Fahrenheit, it cruises coastal environments and depths as great as 3,500 feet, and it ranges across territories as diverse as the offshore waters of South Africa, Oceania, and both South and North America. Moving its 12-to-20-foot-long mass through the water with oiled ease, it satisfies a carnivorous appetite with a bite force of up to 4,000 pounds per square inch, feeding on seals, turtles, seabirds, dolphins, porpoises, tuna, and other sharks—in short, almost everything below its lofty position on the food chain. And perhaps more frightening, among marine animals this predator is ranked number one in the number of recorded attacks on humans.

The great white has earned its rating as a heavyweight predator by using an estimated 250 teeth to tear prey into plate-sized chunks that it swallows whole. It can ingest eight seals during any one mealtime. Compared to the grandiose teeth of Megalodon (on average 3 to 5 inches long and up to 7 inches), the great white's teeth are considerably smaller (1 to 2 inches long) and edged by a series of very pronounced serrations.

That sector of Green Mill Run that flows between Green Springs Park and Elm Street Park, in Greenville, is a noted North Carolina location for finding great white teeth. The high-booted wader may discover crocodile or mosasaur teeth, pieces of whale bone, and assorted javelin-shaped belemnite fossils, but the prize of a patient search is a tooth of the great white, dating to the Pliocene Epoch. The best time to visit is several days after a rain, once the water depth has returned to normal. Successful fossil hunters most often use the screening method to examine shovel loads of stream gravel and work carefully since the creek banks above water level are private property.

In 2001 the National Film Registry chose to have *Jaws* preserved in the Library of Congress on the basis of its cultural, historic, and aesthetic importance. Similar adjectives might well be used to describe a tooth collection of the extant great white, arguably the best known of a long heritage of sharks that dates back to the deep time of the Ordovician Period, hundreds of millions of years ago.

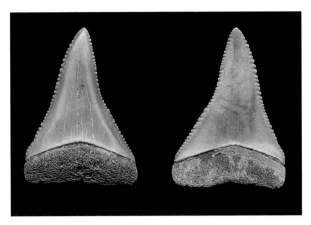

Mesial (left) and distal (right) view of a typical 1.5-inch-long great white shark tooth recovered from Green Mill Run. —Courtesy of Jayson Kowinsky

Green Mill Run fossils, such as this 5-inch-long whale tooth, have blackened with time, a coloration produced by a chemical reaction between calcium phosphate in the fossil and seawater. Other colorations are attributed to exposure to groundwater after the fossil was buried. Penny for scale. —Courtesy of George W. Powell Jr.

Green Mill Run fossils include teeth of the great white (serrated types) and other species of shark, orange belemnites, and a miscellaneous lot of bone detritus from crocodiles, whales, porpoises, mosasaurs, and the occasional dinosaur. —Courtesy of Jayson Kowinsky

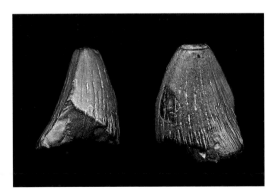

Fossils at Green Mill Run are often reworked, meaning they are very worn and even broken, as seen in these 1-inch-long mosasaur specimens. —Courtesy of Jayson Kowinsky

66. Heritage Center and State Museum, North Dakota

46° 49' 10" North, 100° 46' 48" West

The "big six" tyrants battled for supremacy around a massive seaway.

About 70 million years ago, North America was experiencing the final effects of an evolutionary arms race that had been underway for more than 150 million years. Soon the great Cretaceous extinction event would set the stage for the modern world, but in the meantime the majority of the Peace Garden State was awash in the warm, tropical waters of the Western Interior Seaway, a massive marine invasion that extended 600 miles from the embryonic Rocky Mountains to the aging Appalachians, and 3,000 miles from the Gulf of Mexico to the Arctic Ocean.

Fossils that have been discovered in sedimentary rocks deposited throughout North Dakota during the Cretaceous Period are evidence it was home to a second-to-none menagerie of marine and terrestrial organisms, as exemplified by a "big six" population of tyrants: *Hesperornis*, *Archelon*, and *Xiphactinus* ruled the waters of the seaway through intimidation, while onshore *Tyrannosaurus*, *Triceratops*, and *Edmontosaurus* bullied terrestrial denizens.

Powered by massive hind legs and lobe-shaped toes, *Hesperornis* was a fast, streamlined, fish-eating diver possessed with Olympian agility. Armed with an elongated tooth-lined beak, this 5-foot-long flightless bird was no slouch underwater, but when it ventured onshore to breed and nest, dinosaurs were a constant threat. In contrast, the slow-moving, long-distance-swimming omnivore *Archelon* lumbered through the water casually sweeping up carrion, jellyfish, and plants with its powerful, toothless beak accentuated by a pronounced overbite. Crowned the largest prehistoric turtle that ever lived, this 2.5-ton, 15-foot-long behemoth was, however, quite vulnerable because it couldn't withdraw its head and flippers inside its leathery shell. With a head the size of a grizzly bear and a muzzle like a bulldog's, the 20-foot-long, 0.5-ton *Xiphactinus* was the largest bony fish to troll the Western Interior Seaway. Capable of speeds of 35 miles per hour, it could swallow whole prey up to 6 feet long, including an occasional and unsuspecting *Hesperornis*.

Textbooks devoted to the event of the Maastrichtian Age of the Cretaceous Period, the 6-million-year-long segment of time that defines the final chapter of the Mesozoic Era, emphasize two preeminent terrestrial reptiles. Well-known to every devotee of the Age of Reptiles, the iconic, terrestrial *Triceratops* and *Tyrannosaurus* shared characteristics of size, bearing, and reputation but differed significantly in their eating habits: the three-horned *Triceratops* was a practicing vegetarian, whereas *T. rex* was addicted to flesh.

In 2007, however, a front-page dinosaur discovery in the Hell Creek Formation of North Dakota made paleontologists' hearts beat with renewed vigor. Dubbed "Dakota" and suspected to be a species of the plant-eating *Edmontosaurus* genus of duck-billed reptiles, the collection of skeletal fossils includes examples of ligaments, tendons, skin, and even remnants of internal organs. Fossilized impressions of dinosaur skin were first found more than a century ago, but Dakota is enveloped by 2 inches of three-dimensional, fossilized epidermis, accentuated by well-preserved scales, that overlies a tough, bone-protecting inner sheath. Computer analyses confirmed that this dinosaur "mummy" had the potential to outrun *T. rex* by a lifesaving 8 miles per hour.

The retreat of the Western Interior Seaway southward from North Dakota spelled an end to the Cretaceous Period and extinction to much of its faunal life. Exemplar exhibits of these game-changing, end-of-an-era organisms—organized around skeletal adaptations of the "big six" tyrants—are on display in the Adaptation Gallery of the North Dakota Heritage Center and State Museum (612 East Boulevard Avenue) in Bismarck, where a "back to the Cretaceous" excursion is a captivating and spellbinding deep-time adventure.

Capable of traveling at high speeds for short distances, and guided by a keen sense of smell, T. rex fed on many kinds of its dino brethren, using its powerful jaws and sharp teeth to tear off as much as 500 pounds of flesh in one gulp. —Photo by Brian R. Austin, courtesy of the State Historical Society of North Dakota

Armed with an upturned jaw lined with giant fang-like teeth, Xiphactinus *trolled the surface waters of the Western Interior Seaway.* —Photo by Brian R. Austin, courtesy of the State Historical Society of North Dakota

Dakota, the dinosaur "mummy," is one of the most important dinosaur specimens ever found because of its preserved skin envelope, complete with what appear to be cell-like structures. Inch ruler for scale. —Photo by Brian R. Austin, courtesy of the State Historical Society of North Dakota

Triceratops *employed its imposing horns to scare off predators and defend itself against them, as well as for sexual display during mating season. Its skull is one of the largest of any land animal, past or present.* —Photo by Brian R. Austin, courtesy of the State Historical Society of North Dakota

67. Pembina Gorge, North Dakota

48° 55′ 01″ North, 98° 03′ 22″ West

The mosasaur, a distant relative of the Komodo dragon, terrorized the seas of the Peace Garden State.

The Rendezvous Region of northeastern North Dakota, named in memory of the annual frontier-era "rendezvous" gabfests that brought trappers, Indians, and traders together to exchange gossip and barter and let off a little steam, is a land of colorful description. Three different biogeographical provinces—grassland prairie, deciduous forest, and coniferous forest—merge in contrasting harmony to form a 13,000-acre mosaic of wetlands and woodlands that is home to a diverse fauna, including an unexpected herd of elk.

During the Cretaceous Period, long before humans of any kind set foot in the region, one-third of the landmass of Earth was inundated by epicontinental seas. In North America this great flood, named the Western Interior Seaway, extended from the Gulf of Mexico to the Arctic, and from the Rocky Mountains to the Appalachians. The waters teemed with a census of animals that differed significantly from those that populated seas of the Paleozoic Era. No longer did the seafloor harbor extensive communities of horn corals, brachiopods, crinoids, trilobites, and bryozoans. Predatory organisms now included gargantuan turtles, giant squid, and, perhaps most spectacular and ferocious of all, mosasaurs. These extinct 30-to-40-foot-long modified land reptiles are ancestors of the modern-day Komodo dragons of Indonesia.

The story of the mosasaur is an unusual chapter in the history of paleontology. Known also as the *T. rex* of the marine world, these predatory marauders ruled the seas during a 20-million-year reign of terror made possible by their apparent ability to outmaneuver all other forms of life in their hunt for food. They employed modified flipper-like limbs for supple, sinuous motion through the water. At the end of the Cretaceous Period, these streamlined, large-jawed, lizard-like, agile, alpha carnivores became extinct, along with the dinosaurs.

Skeletal remains of mosasaurs, in association with a diverse assemblage of invertebrate fossils, are distributed throughout the Pierre Shale, the Cretaceous-age sedimentary rock deposited in the Western Interior Seaway. They might have remained buried had it not been for the ice age. Several million years ago a looming, 1-mile-thick wall of ice slowly ground its way across the northern plains, tearing, gouging, and smashing everything and anything in its path. In time this continental glacier reached its most southern point and then began to retreat, a victim of global warming. In its wake surging meltwater sculpted Pembina Gorge and exposed the Cretaceous-age fossils. Today, the meandering Pembina River continues to deepen the gorge and expose new evidence of mosasaur fossils.

Due to the fragile and disarticulated nature of mosasaur fossils, paleontologists generally remove them using extraction techniques unavailable to the weekend fossil enthusiast. In recognition of this difficulty, paleontologists from the North Dakota Geological Survey conduct summer programs for families and children at an excavation site in Pembina Gorge, where 80-million-year-old remains of mosasaurs, birds, fish, and 15-foot-long squid have been found. Half-day and full-day experiences here, and at other excavation sites, are available through a public fossil-dig program (https://www.dmr.nd.gov/ndfossil/digs).

A restored skeleton of a 23-foot-long mosasaur, along with fossils of other marine organisms of the same time period, is on display in the Adaptation Gallery in the North Dakota Heritage Center and State Museum (612 East Boulevard Avenue) in Bismarck. Though not all of the mosasaur bones found at Pembina Gorge have been identified, paleontologists have determined that those of the Bismarck skeleton represent a new species found nowhere else in the world.

The Western Interior Seaway, considered one of the greatest epicontinental seas of all time, created a placid, tropical environment in central North America throughout the Cretaceous Period.

A partially exhumed, 8-inch-wide mosasaur vertebra lies on a foundation of Pierre Shale in Pembina Gorge. A second vertebra (upper left) remains in its original, untouched position. —Courtesy of the North Dakota Geological Survey

Once discovered, the specific size, shape, and orientation of Pembina Gorge fossils are recorded using a grid-mapping process. —Courtesy of the North Dakota Geological Survey

Due to the fragile nature of many of the Pembina Gorge mosasaur bones, fossil specimens are encased in plaster jackets (white objects, bottom center) prior to transportation to the laboratory, where paleontologists will examine them. —Courtesy of the North Dakota Geological Survey

68. Caesar Creek State Park, Ohio

39° 28′ 48″ North, 84° 03′ 26″ West

Fossils that inspired the founding of the Cincinnati School of Paleontology.

Cincinnati, Queen City of the West, has experienced several unusual moments of fame. Cue the drumroll: its onetime role as the chief hog-packing center of the country; its role in the Underground Railroad, the shadowy system of escape routes that brought freedom to many slaves; and as the home base of the Cincinnati Red Stockings, the very first professional baseball team. Lesser known is its association with the Cincinnati School of Paleontology.

Organized in the 1870s by a group of hobbyists with a common interest in the fossils found in the rocks of the greater Cincinnati area, the "school" flourished until the last member of the founding group died in 1961. Educated in the fields of law, medicine, and physical geology, these publishing amateurs viewed the study of fossils an avocation rather than a profession. As such, they were roundly and routinely ostracized by credentialed East Coast–based paleontologists who plied their trade within the hallowed halls of academia and government.

Undaunted, the members of the school continued to scour the quarries and outcrops of the greater Cincinnati area in search of ever more fossils that were exclusive to the region's sedimentary strata. Known as the Cincinnatian Series, these rocks represent the youngest of the three epochs that constitute the Ordovician Period in the United States.

The greater Cincinnati area, defined by any invertebrate paleontologist of worth as a 50-mile-diameter circle enclosing the tristate region of Ohio, Kentucky, and Indiana, is globally famous for the quality, diversity, and quantity of its invertebrate fossils, specimens of which make up the collections of major geologic museums throughout the world. More than seven hundred species of Ordovician life have been discovered in the area within Cincinnatian Series strata, and the quantity of certain species is staggering. It has been said that if all the fossils found in the bedrock underlying Cincinnati were somehow removed, the city would sink below sea level. There is no better place to begin a collection of Late Ordovician fossils than Caesar Creek State Park, about 30 miles southeast of Dayton, where a free-of-charge collecting permit is available at the visitor center.

Four species of coral, nine of bryozoans, and twenty-four of brachiopods can be found in the three limestone and shale formations exposed within the wall of the emergency spillway, the favored collecting site. The prima donna billing, however, belongs to *Isotelus*, the trilobite genus designated Ohio's official state invertebrate fossil in 1985. First cousins to today's crabs, lobsters, scorpions, centipedes, and spiders, trilobites probed the world's oceans for almost 300 million years, from their appearance during the Cambrian Period to their extinction in the Permian Period. Most of them burrowed through muddy seafloor sediment ingesting nutrients, much like a present-day earthworm does as it tunnels through soil.

At least eight different species of trilobites are found in Caesar Creek State Park. On occasion one might discover one in a curled-up state, a position taken at the moment of death, supposedly as a defense posture against predators, but most specimens are fragments. They range in size from that of an ant to more than 18 inches in length. A truly outstanding, 1-foot-long, prone example of *Isotelus*, unearthed during an excavation along the emergency spillway, is on prominent display at the visitor center.

Extensive rubble piles of Ordovician-age sedimentary rock litter the fossil-collecting grounds at Caesar Creek State Park.

Studies of museum-quality Isotelus, *such as this 1-foot-long specimen, indicate this predator hunted worms and other soft-bodied benthic animals in the mud of the Ordovician seas.* —Courtesy of the Ohio Department of Natural Resources, Division of Geological Survey

These bivalved brachiopods lived and died in shallow water, as evidenced by the presence of disconnected valves lying with their convex side up, in a stable position. This state of fossil preservation is generally associated with wave and tide forces strong enough to flip over convex-side-down specimens. Nickel for scale.

Brachiopods are sedentary, antisocial creatures dwelling in a bivalve housing that must be heavy and thick to survive in shallow, agitated water. Rafinesquina, Platystrophia, *and* Hebertella *are the types most commonly found at Caesar Creek State Park.* —Courtesy of Kendall Hauer, Miami University

69. Sylvania Fossil Park, Ohio

41° 42′ 46″ North, 83° 44′ 52″ West

**Fossils of creatures that thrived until a little-known—
and controversial—mass extinction event brought devastating change.**

The history of life on Earth is punctuated by many recognized *mass extinctions*, events that significantly exceeded background extinction rates and affected a wide variety of life. A handful were particularly severe, such as those occurring at the end of the Ordovician, Permian, Triassic, and Cretaceous Periods. Consensus exists among paleontologists that each of these four lasted between one and five million years. The timing and duration of a fifth mass extinction, commonly referred to as the Late Devonian event, is subject to considerable controversy. Paleontologists recognize that anywhere from two to ten specific extinction episodes, varying in importance and spread over a total time period of some 15 to 25 million years, were associated with the Late Devonian mass extinction. Two merit special recognition: the Kellwasser event marks the initiation of the final phase of the Devonian Period, and the Hangenberg event brought the curtain down on the Devonian.

By Late Devonian time, a plethora of plants, forests of primitive trees, and a potpourri of swarming insects blanketed the land, and invading creepy crawlers—centipedes, scorpions, snails, nematodes, and millipedes—were fighting to establish territories for themselves. However, multiple sharks and ray-fins, and the armored *Dunkleosteus*, a 35-foot-long giant capable of delivering a bite force of 80,000 pounds per square inch—130 times the bite force of a modern shark—justifies labeling this geologic period the Age of Fishes.

The early stages of the assembly of the supercontinent Pangaea were underway, and the landmass that would become North America was slowly drifting north across the latitude of 10 degrees south. Surrounding oceanic waters teemed with invertebrate life; brachiopods, corals, cephalopods, gastropods, trilobites, graptolites, and the often-rooted animal known to fossil enthusiasts as "sea lilies" inhabited waters warmed and oxygenated by tropical breezes.

Then, the one-two punch came—disastrous, far-reaching, and complete and irreversible for many forms of life. Most of the damage occurred during the Kellwasser event, but the Hangenberg event, believed to be about 70 percent as severe, was also of momentous consequence. Educated opinions disagree as to the overall extent of the Late Devonian extinction event, but at the genus level it's estimated that 50 to 60 percent of Earth's life went extinct, while at the species level it was 70 to 75 percent. A wide swath of marine life was affected: reef-building organisms were almost annihilated; some 90 percent of floating plants died; one-third of bryozoan genera were lost; corals experienced a crushing decline; and the biomass loss of brachiopods, until then the most dominant of shellfish, can best be described as devastating. The causes of the Devonian extinction are still under heated discussion, with suggestions ranging from climate change to multiple meteorite impacts to the effects of volcanism.

An ideal site to examine the remains of life that existed immediately before the Devonian extinction event is Fossil Park, located in Sylvania. Billed as one of only two prime Devonian fossil locales on the entire planet, this 5-acre site is open to the public and periodically restocked with truckloads of fossil-rich Silica Formation rock, harvested from a nearby quarry. Because this formation was deposited immediately prior to the onset of the Kellwasser event, an easily collected assemblage of common brachiopods, numerous crinoids and corals, and segments of trilobites is indicative of the way it was in those distant days when devastating change lurked just beyond the horizon.

Ranging from 2 to 3 inches long, Megastrophia *is one of the largest fossils found at Fossil Park. Size and the presence of thin, closely spaced, radiating ribs are the distinctive features used to identify this brachiopod.* —Courtesy of Kendall Hauer, Miami University

Phacops *is prized by professional and amateur alike, but generally only bits and pieces of this trilobite are found at Fossil Park. This species, recognized by its large eyes, is named* rana *in reference to the bulging eyes of the common frog (*Rana temporaria*).* —Courtesy of Kendall Hauer, Miami University

Mucrospirifer *brachiopods (pedicle valve on left, brachial valve on right) are fairly abundant in the debris piles at Fossil Park. These 1-to-2-inch-wide, exquisitely preserved, 370-to-400-million-year-old fossils easily weather out of the soft, gray shale of the Silica Formation and clean up nicely with an electric abrasive tool using compressed air.* —Courtesy of Kendall Hauer, Miami University

70. Arbuckle Anticline, Oklahoma

34° 24′ 20″ North, 96° 57′ 02″ West

Ordovician-age brachiopods and companion fossils are present in both roadside and showroom contexts.

Named after Colonel Matthew Arbuckle, commanding officer of the Seventh Infantry Regiment who, in 1824, established the first US Army fort in what was then the Indian Territory, the Arbuckle Anticline is one of three similar northwest-southeast-oriented structures that collectively form the Arbuckle Mountains in south-central Oklahoma. The history of this now-denuded highland, the oldest known geographic feature between the Appalachian and the Rocky Mountains, began during the Cambrian Period some 500 million years ago with the development of a rift, a tectonic powerhouse with the dual effect of rupturing Earth's crust and bifurcating the ancestral North American continent. After 100 million years of activity the rift became dormant, but not until its depths had been flooded with massive layers of volcanic rock. The cooling and contracting of this igneous rock basement formed a basin-like depression that an extensive shallow sea inundated. For the remainder of the Paleozoic Era, calcium-rich remains of marine organisms and sediment eroded from the surrounding headlands filled the basin, forming sequences of fossil-bearing limestone and shale in excess of 14,000 feet thick in some regions.

Toward the end of the Paleozoic Era the region was subjected to a second episode of tectonic disruption and destruction caused by compressional forces associated with the assembly of the supercontinent Pangaea. Flat-lying sedimentary rocks were thrust upward, folded, and faulted into the Arbuckle Anticline—and the mountains of the same name. Today, after millions of years, the once lofty highlands have been eroded from heights that may have exceeded 15,000 feet to a moderately dissected, low-lying plateau with its highest point poised a mere 1,415 feet above sea level. Renowned for their beauty, the Arbuckles are valued as a treasure trove of natural resources: limestone, dolomite, granite, glass sand, iron ore, lead, zinc, tar sands, oil, gas—and fossils.

Strata ranging from Cambrian through Pennsylvanian age are beautifully exposed along I-35 and the surrounding north-south highway network between the communities of Davis and Ardmore. Two-thirds of the thickness of the folded and faulted rock is composed of Ordovician-age carbonate strata, the most significant being limestone layers of the Bromide Formation. Age dated to approximately 460 million years ago, Bromide fossils are significant because they represent fauna with life spans bookended between the Great Ordovician Biodiversification Event, a big bang–like explosion in the variety of marine life, and the Ordovician extinction event, the first of five such life-altering incidents of geologic history.

Fossils found in the Bromide Formation include a variety of corals, graptolites, bryozoans, mollusks, trilobites, and brachiopods. Trilobite and brachiopod fossils, the latter also known as lampshells, are present in the Bromide strata exposed in the Arbuckle Anticline roadcut along the east side of US 177, 7 miles south of the southern limits of Sulphur. These rocks lie within the north flank of the anticline, as evidenced by the northerly plunge of the strata. More than eighty brachiopod genera have been reported from the Bromide Formation.

Enthusiasts wishing to delve further into the world of Arbuckle fossils should arrange for a weekday visit to the showrooms of Geological Enterprises (308 Stolfa Street SE), in Ardmore. Billed as having the largest selection of fossils available anywhere, this facility has supplied Smithsonian-quality specimens to museums and classrooms throughout the world for six decades. Their collection of Arbuckle Mountains specimens is especially impressive, many laboriously unearthed and finished into exquisite three-dimensional presentations.

Much work remains before the full beauty of this Strophomena *brachiopod (below quarter) is extracted from the Bromide Formation limestone collected at the Arbuckle Anticline roadcut.* —Courtesy of Virginia Ford, College Station, Texas

Clusters of Oxoplecia gouldi *brachiopods are occasionally found in the Bromide Formation, but never in this form. Museum-quality displays such as this require multiple hours of patient preparation by experts schooled in the art of exhuming fossils, by mechanical and chemical processes, from their deep-time burial. The brachiopod in the upper right position is 2 inches wide.*
—Courtesy of Geological Enterprises, Inc., Ardmore, Oklahoma

A baker's dozen of 2-inch-long, chocolate-brown Homotelus bromidensis *trilobites nestled in the muted yellow matrix of the Bromide Formation.* —Courtesy of Geological Enterprises, Inc., Ardmore, Oklahoma

Ordovician-age fossils are present across the full extent of the Arbuckle Anticline outcrop along US 177, but the best zone is reportedly confined to the yellowish rock, seen in the center of this view. —Courtesy of Virginia Ford, College Station, Texas

71. John Day Fossil Beds National Monument, Oregon

44° 33′ 09″ North, 119° 38′ 46″ West

One of the longest and most complete records of evolutionary change in the world.

There indeed was gold in "them thar hills" around Canyon City, Oregon, and the presence of troops of the US Cavalry gave assurance that wagonloads of the precious ore could be safely delivered to shipping docks on the Columbia River. Operations progressed smoothly until the day in 1862 when soldiers discovered fossils along the route from mine to ship, and inquiries of provenance became as relevant as concerns of security. Fossils of plant and terrestrial vertebrates were passed along to Thomas Condon, a local minister and self-taught geologist, for preliminary identification. Recognizing their importance, Condon sent samples east to several universities for further analyses. Soon a second rush west was underway along the Oregon Trail as paleontologists joined wannabe gold prospectors in the pursuit of fame and fortune.

Thomas Condon died in 1907, leaving a record of accomplishments, the highlights being named the first state geologist of Oregon and the first professor of geology at the University of Oregon. His pioneering investigations form the core of a catalogue of information documenting 40 million years of paleoclimate and evolutionary change that took place in the western hemisphere during the Paleogene and Neogene Periods. Today, the Thomas Condon Paleontology Center is the central attraction within the 14,000-acre parcel of fossil-bearing lands known as John Day Fossil Beds National Monument.

Throughout the last quarter of the nineteenth century the many expeditions that probed the richness of the John Day Fossil Beds focused on bones. Any kind of bone would do, but during this era the most highly prized specimens were intact skulls with teeth. Wagonloads of vertebrate remains were shipped to museum curators throughout the country to be assembled into charismatic exhibits designed solely to dazzle and mystify the viewing public.

With the dawning of the twentieth century, a new generation of scientists began to recognize the fossil beds not as a simple cache of ancient bones but rather as a colossal window into the past, perhaps the most diverse and best-preserved record of Cenozoic terrestrial life anywhere on Earth. Bones were still important, but now field collections included any and all evidence—soil and rock samples, plant fossils, and microorganisms—that might lead to a systematic understanding of life during the Age of Mammals. In turn, museums enhanced their exhibits to include educational panoramas that documented changes from a lush, semitropical "greenhouse" environment 44 million years ago to a global "icehouse" climate that preceded the onset of the Pleistocene ice age.

Fossil collecting is not allowed in the monument, but eight large and dynamic murals are on display at the Thomas Condon Paleontology Center, 2 miles north of the intersection of US 26 and Oregon 19, northwest of Dayville. Among the more than 45,000 catalogued specimens representing 2,100 species are those related to primitive horses and rhinoceroses, primitive crocodiles, evolving camel-like grazers, carnivorous mammals, burrowing rodents, saber-toothed animals of varied sizes, and the largest and most predatory dog of its time. Exhibits also include more than 300 plant species and miscellaneous fish, amphibian, bird, and insect fossils.

John Day Fossil Beds National Monument is well deserving of its reputation as home to one of the richest fossil beds on Earth. If a more complete paleontological record of the years following the catastrophic extinction of the dinosaurs exists anywhere in the world, it has yet to be discovered.

The 50-degree attack angle of the jaws of this giant, 3-foot-long Archaeotherium *skull is proof this piglike scavenger lived competitively with other Eocene animals, such as saber-toothed cats and warthog-like creatures as tall as bison.* —Courtesy of Joshua Samuels

This 7-inch-wide specimen offers a representative sampling of the 175 species of fossil fruits and nuts and at least 75 species of wood—more than any other petrified wood locality in the world—that establish the John Day Fossil Beds as a world-class locale. —Courtesy of Joshua Samuels

The 2011 discovery of these 0.8-inch-long teeth provides evidence that beavers crossed the Bering Land Bridge at least 7 million years ago, not 5 million as previously thought. It seems appropriate they were found in John Day strata as Oregon carries the nickname the Beaver State. —Courtesy of Joshua Samuels

Exposed to sunlight for the first time in some 40 million years, this 1-foot-long fossil of Stylemys, *an extinct, herbivorous dryland tortoise, appears ready for a museum exhibit. Its dome-shaped shells are relatively common in John Day strata, but heads are extremely rare.* —Courtesy of Joshua Samuels

72. Lava Cast Forest, Oregon

43° 49′ 02″ North, 121° 17′ 18″ West

Impressions of a millennia-old forest preserved as molds.

During the summer months in the years of the early 2010s the forests of the west coast were devastated by wildfire, the result of a drought that had plagued the region for years. The Rim Fire of 2013—the third largest in California history—alone burned across 400 square miles of Sierra Nevada countryside for nine weeks. The combination of warm temperatures, strong winds, dry weather, minimal rainfall, and a spark can ignite a blaze capable of moving up to 15 miles per hour, destroying everything in its path: trees, brush, homes, and, occasionally, entire communities. Conventional analyses indicate 80 percent of wildfires are caused by people (a not-quite-snuffed-out campfire, a downed power line, or arson), with atmospheric conditions (lightning and such) responsible for the remaining 20 percent. Geologic records, however, show that forests can be incinerated by other means. Lava Cast Forest, a 5-square-mile section of Newberry National Volcanic Monument, is a case in point.

About 6,000 to 7,000 years ago western Oregon appeared much as it does today, carpeted by forests of white fir, lodgepole and ponderosa pines, and a groundcover of grasses and shrubs. The Cascade Range, constructed on an igneous foundation of thirteen supersized volcanoes and numerous lesser eruptive centers distributed over 700 miles, was becoming, with every passing century, a force ever more capable of catastrophically rearranging the scenery of California, Oregon, and Washington.

Newberry Volcano, about 20 miles southeast of Bend, is one of these less significant volcanic centers of the Cascade Range. Even still, it has a volume of 80 cubic miles and a diameter of 20 miles, and is one of the largest assemblages of cinder cones, lava domes, lava flows, and fissure vents in the world. Numerous eruptions spread over hundreds of thousands of years constitute the bulk of its geologic history. One such eruption, occurring around 4,500 BC, would be generally unrecognizable today had it not left in its wake features of a most unusual nature: fossil tree molds.

Molten rock that erupted onto Earth's surface, at temperatures approaching 2,000 degrees Fahrenheit, flowed around and between the trunks of the Newberry forest. Before woody fibers could spontaneously burst into flame, a crust of fine-grained lava crystalized around each tree, the result of cooling temperatures induced by the interaction of tree moisture and molten rock. In short order the then-dehydrated trees burned to ash, leaving molds replicating the bark-like impressions of a forever vanished forest.

Tree molds in Lava Cast Forest can be classed as *trace fossils*, the indirect evidence, such as tracks, tubes, trails, casts, and molds, of former life. In this case "lava cast" is a misnomer, because a mold forms around an object, whereas a cast fills a mold. Several different types of molds can be seen along the 1-mile-long trail that traverses Lava Cast Forest: some molds horizontal in attitude, a threesome connected to a parent root system, and twins growing side by side. A few molds extend several feet below the surface.

A stroll along the paved Lava Cast Forest Trail, with purple penstemon and red-orange paintbrush in full bloom, is an experience not found anywhere else in America. Whether called *casts* or *molds*, the fossil voids of this walk through time remain unique in both appearance and manner of creation.

Many molds in Lava Cast Forest have a dramatic fine-grained crust that was created when tree moisture induced the invasive lava to rapidly crystallize. Pen for scale.

Side-by-side fossil tree molds are present-day evidence that a once-flourishing forest was incinerated by lava.

Lava trickling and dripping into the space between the original mold and charred wood created patterns commonly mistaken for bark impressions. Pen for scale.

73. Montour Preserve, Pennsylvania

41° 06′ 38″ North, 76° 38′ 50″ West

Representative fossils of life at the brink of disaster.

The annals of geologic history are commonly divided into segments that end with Earth-rending episodes, such as earthquakes, volcanic eruptions, and meteorite impacts. Like the most mind-wrenching chapters of a best-selling murder mystery, which go from mild to wild with the turn of a page, geologic episodes of normalcy are abruptly terminated by consequences of disaster. The hillside setting of Fossil Pit, located on the grounds of the Montour Preserve (700 Preserve Road) in Danville, represents the mild point of a chapter right before the plot line accelerates. Here exposures of shale and siltstone of the Mahantango Formation reflect yet another calm, placid, unexciting day of the Devonian Period, yet doomsday lay immediately beyond the geologic horizon.

Two geographic vistas dominated the landscape of the United States 395 million years ago: the Acadian Mountains in the east and the Kaskaskia Sea to the west. Sediment-saturated rivers draining the highlands deposited volumes of mud and silt throughout Maryland, New Jersey, New York, and Pennsylvania—sediment that would be compacted to form the Mahantango Formation. The Kaskaskia Sea was shallow (from 60 to 200 feet), warm, and highly oxygenated, conditions suggestive of a body of water flush with life. If the number of fossils collected over the years at the Montour Preserve's Fossil Pit is any indication, two-thirds of that life was composed of brachiopods, thumb-sized animals that secreted an exoskeleton made up of two valves. The common appearance of brachiopod valves in shale indicates the organisms thrived in muddy water not normally disturbed by strong currents or storm waves; shale is composed of fine sediment that settles out in calm water. The remainder of the specimens found at Fossil Pit, in decreasing order of abundance, includes corals, cephalopods, pelecypods, gastropods, crinoids, bryozoans, and trilobites. This fossil richness is attributed to the presence of a reef system that trended northeast-southwest through Pennsylvania during Middle Devonian time. Present-day reefs, often called "rain forests of the sea," are organic structures that provide three-dimensional habitats for a full 25 percent of all ocean species. Their presence, now and in the geologic past, serves as evidence that this segment of the Devonian Period was a time of organic prosperity.

In Late Devonian time the sun rose on a very different scene. Within a mere 5 to 20 million years—a very short period of geologic time—the worldwide food chain was turned topsy-turvy and consumed by evolutionary disorder. More than 75 percent of all species on Earth became extinct, a loss that included even *Dunkleosteus*, the armored, 35-foot-long alpha predator that roamed the Kaskaskia Sea, ever ready to apply its bite force of 80,000 pounds per square inch on any available prey. The suggested causes of this second of the five great extinction episodes is subject to debate. Evidence supporting global cooling, sea-level decline, sea-level rise, meteorite impact, reduced oxygen level in the ocean, and changes in atmospheric chemistry have all been posited and embraced to varying degrees. Whatever the reason, or reasons, it was a time that forcefully reorganized the character of life on Earth on a major scale.

The fossils found on the talus slopes of Fossil Pit differ little in appearance from those discovered at other Devonian sites. Their true significance lies in the fact that they represent a state of becalmed life that evolutionary chaos would soon transform. They were harbingers of disaster, death, and, ultimately, irreversible loss.

You can travel back in time and search for fossils along the seafloor of the Kaskaskia Sea, but be sure to bring a hammer, chisel, and safety glasses, because the specimens at Fossil Pit are well entombed in rock. Hammer for scale. —Courtesy of the Montour Preserve

It takes an eagle eye to find a trilobite at Fossil Pit. They are rare, commonly coiled up like this 1-inch-wide specimen, heavily weathered, and decidedly untrilobite-like in appearance. —Courtesy of the Montour Preserve

Two-thirds of the Fossil Pit specimens are brachiopods. Occasionally a whole fossil is unearthed, but commonly they are found as imprints, such as this hazelnut-sized specimen of the Mediospirifer *genus, showing radiating ornamentation.* —Courtesy of the Montour Preserve

Cone-shaped horn corals were abundant from Middle Ordovician to late Permian time. This 1-inch-long example was discovered lying loose at Fossil Pit. —Courtesy of the Montour Preserve

74. Red Hill, Pennsylvania

41° 20′ 40″ North, 77° 40′ 50″ West

Marine life blazes a fin-to-limb trail onto land to escape a fish-eat-fish world.

The whole of geologic history can be divided into two eons, the Precambrian and the Phanerozoic. The Phanerozoic is bookended by the Cambrian explosion, that quintessential episode that began 541 million years ago when every major animal phyla burst onto the evolutionary scene, and the Quaternary Period, beginning 2.6 million years ago, continuing today, and often referred to as the ice age and the Age of Man. There are also many intermediate Phanerozoic events of note, such as the extinction of some 95 percent of oceanic species at the end of Permian time, the demise of the dinosaurs at the end of the Cretaceous Period, and the transition of aquatic life to land. Though not as widely known, the latter event was not necessarily less significant.

The Devonian, 419 to 359 million years ago, was the Age of Fishes. Jawless types were learning to coexist with their acanthodian cousins, fish with jaws and cartilage skeletons, including the first true sharks and the armored placoderms. Groups supported by skeletons made of bone rather than cartilage—ray-fins and lobe-fins—also occupied the Devonian stage. The ray-fins would evolve to become the dominant fish of the contemporary world, but it is the lobe-fin that perhaps forever changed life on Earth, giving rise to four-legged tetrapods that fostered the transition from water to land.

The Devonian was an opportune time for any fish wishing to stake out a new lifestyle on land. Globally, terra firma was distributed between two main masses: Gondwanaland in the southern hemisphere and Laurasia straddling the equator to the north. The continental-shelf waters of Laurasia, charged with oxygen and heavily nurtured by sunlight, were incubators for life-forms seeking new evolutionary niches.

Tetrapod fossils have long been known from regions as diverse as Scotland, Latvia, Russia, China, Greenland, and Australia, but they remained undiscovered in North America until 1993, when bones were found at Red Hill, along Pennsylvania 120, 2 miles northwest of Hyner. The extensive fossil list of Red Hill, including sharks, ray-fins, lobe-fins, insects, and a variety of vascular plants, constitutes one of the premier assemblages of Devonian plant and vertebrate remains in the world. The star attractions, however, are *Hyneria*, a giant, carnivorous lobe-fin, and *Hynerpeton*, a four-legged creeping critter—both named after the nearby hamlet of Hyner.

The shoulder bones of *Hynerpeton*, extracted from the Devonian-age Catskill Formation at Red Hill, are suggestive of a robust, 6-foot-long, very muscular amphibian that frequented river, estuary, and terrestrial landscapes. Large muscle scars indicate it could navigate forward and backward with equal ease, either searching for a mate or foraging to satisfy its supposed diet of fish and insects. This find is dubbed the oldest tetrapod fossil found in North America.

Hyneria fossils are also found at the Red Hill exposure, but they are contained in Catskill Formation strata deposited in nearshore and floodplain environments. Some paleontologists theorize that these monstrous predators used their powerful fins to haul themselves onshore for short periods of time when—and where—more opportunistic conditions prevailed.

The story of Red Hill is one of evolving lobe-fin fish learning to strengthen their limbs to legs, crawl onto land, and take a gulp of fresh air, setting the stage for the development of land-dwelling animals. As *Hyneria* the lobe-fin evolved to *Hynerpeton* the tetrapod, faunal life on Earth occupied, for the first time, both oceanic and continental enviroments. A visit to Red Hill can be combined with a stop at the Chapman Township Municipal Building (196 Main Street), in nearby North Bend, where fossils of *Hynerpeton* and others are on public display.

Measuring more than 1.5 inches in diameter, Hyneria fish scales are indicative of the overall size of these gargantuan predators. Extrapolations based on skull and jaw fragments suggest they reached 10 to 12 feet in length and weighed as much as 2 tons. Ruler for scale. —Courtesy of Jayson Kowinsky

Some of the Hyneria teeth collected at Red Hill were sheared off in several directions. Paleontologists have suggested this less-than-perfect dental condition was caused by tough and prolonged combat with hard-boned prey. Ruler for scale. —Courtesy of Jayson Kowinsky

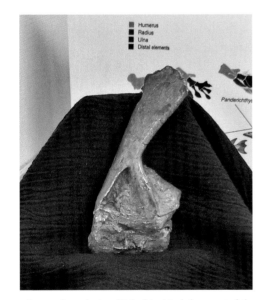

Some paleontologists think this 4-inch-long cast of the left shoulder bone of Hynerpeton, on display at the Chapman Township Municipal Building, is evidence that this early tetrapod linked the fins of fish and the legs and arms of amphibians. —Courtesy of Jayson Kowinsky

The Late Devonian exposure at Red Hill consists of a high vertical wall that is conducive to infrequent rockslides. The wide berm allows for off-road parking, but visitors are advised to wear a hard hat, stay well away from the road, and be constantly aware of vehicles traveling in both directions.

75. Corys Lane, Rhode Island

41° 36′ 04″ North, 71° 16′ 36″ West

An unusual opportunity to collect primitive forest fossils from metamorphic rocks.

The chance of finding a fossil site of great significance in Rhode Island is directly related to the size of the state—in a word, it's small. Several episodes of mountain building and subsequent periods of erosion have, in general, erased any evidence of ancient faunal life that once existed. Pennsylvania-age strata in the state, however, are rich in fossil flora, and this treasure trove of leaves and stems is centered on wave-swept Narragansett Bay.

Narragansett Bay is geologically quite young, shaped when glaciers ravaged the region 17,000 years ago and exposed when they melted and retreated to the north 7,000 years later. Its structural development, however, had begun 350 million years earlier, when Laurentia and Gondwanaland slowly crunched up against each other in the process of forming the huge supercontinent Pangaea. This fender-bender collision of landmasses created a dogleg depression in the crust of Rhode Island that extended north and east into Massachusetts. Rivers flowing from several different directions gradually filled in the 50-mile-long hollow with layers of sandstone, shale, and conglomerate. The area of Narragansett Bay was then located thousands of miles from any coastline, its landscape dotted with low-lying wetlands, marshes, and swamps.

By Late Pennsylvanian time, 300 million years ago, the land had become heavily clothed with scaly trees, seed ferns, club mosses, and giant horsetails. Amphibians, air-breathing mollusks, and a variety of insects inhabited the forests. Trees grew rapidly in the tropical climate, and since there was minimal seasonal change in either humidity or temperature, they lacked distinctive growth rings. Dead vegetation fell into the turbid, murky waters of the forest floor and, insulated against the decaying effects of bacteria and oxygen, formed layers of peat that harbored fossilized leaves, stems, and particles of bark. When the Alleghanian orogeny—the third of the three Paleozoic-age phases of mountain building that created the Appalachian Mountains—reached its crescendo during Permian time, the peat beds were metamorphosed to anthracite, a category of hard coal with a high carbon content.

With a little patience, and at low tide, collectors can gather a representative sampling of Pennsylvanian-age fossils from shoreline anthracite exposures of the Rhode Island Formation at the end of Corys Lane, 1 mile northwest of its intersection with Rhode Island 114 and 1.5 miles due west of Portsmouth. Plant fossils are common. *Calamites* (horsetails), *Pecopteris* (fern leaves), and *Odontopteris* (seed fern) have all been collected from the area. *Calamites* was a tree-sized plant constructed of a hollow bamboo-like trunk with upward-slanted branches whose tips were covered with conifer-like needles. They grew to heights of more than 60 feet. *Pecopteris* resembled a modern palm tree and grew to heights of around 30 feet. *Odontopteris* belonged to an unusual family of extinct plants that reproduced by spreading seeds, rather than spores, upon the forest floor.

Insect fossils at Corys Lane are uncommon, but the 3-inch trace of a mayfly-like bug, believed to be the oldest whole-body impression of a flying insect ever found, was discovered in rocks of similar age in Massachusetts. The outline of its imprint suggests the bug landed in mud, left its body print, and then flew away.

At first glance Corys Lane fossils are hard to see, as they often appear as irregular designs embedded in dark-toned coal. Their black-on-black presence, however, is evidence of a time when primitive, dense, and dark forests thrived in a humid and soggy atmosphere in the region of Narragansett Bay.

Exposures of Rhode Island Formation are easily accessible at Corys Lane during low tide. Plant fossils are confined to black carbonaceous rock adjacent to the shore. —Courtesy of Steve Emma

An important constituent of the Pennsylvanian-age forest of Rhode Island, Odontopteris *is a genus of bushy seed fern so named because of the slanted, toothlike lobes of its leaves—*odont *means "having teeth."* —Courtesy of Steve Emma

A Pecopteris *fossil from Corys Lane that shows a well-defined central stem with attached comblike leaves. This common fern became extinct in the Permian Period.* —Courtesy of Steve Emma

76. Edisto Beach, South Carolina

32° 30′ 18″ North, 80° 17′ 41″ West

Ice age fossils unevenly distributed within random pavements of Holocene shells.

As early as 1847, during a visit to America, Louis Agassiz, the Swiss-born geologist famous for landmark studies of glaciers and extinct fish, made note of the unusual fossil assemblage he found washed up onto Edisto Beach. Seventeen decades later, the mere mention of Edisto fauna is reason for lively discussions among paleontology aficionados interested in the fossil treasures of South Carolina's coastal regions. One group seeks specimens of shell life that lived offshore, whereas others, ardent devotees of vertebrate bones and teeth, gaze ever hopeful to the east to treasure embedded in rock. Sourced from different environments and different periods of time, the Edisto fauna is as contemporary as yesterday and as antique as the ice age

Purists are hesitant to consider contemporary shoreline shells as fossils, but many novices, recalling the liberal definition of a *fossil* as any evidence of former life, consider beachcombing an introduction to the world of paleontology. Edisto Beach is rife with isolated assemblages of shells, each one composed of an amazing variety of colorfully named specimens: banded tulips, Atlantic jackknives, channeled duck clams, slippersnails, wentletraps, angel wings, whelks, apple murexes, and lettered olive, the official shell of the Palmetto State. Some seven hundred species of shell life live offshore of Edisto Beach State Park, and their remains are conveniently accessible from the main office parking lot.

If the inventory at first seems disappointing, wait for the changing tides to add new material to the beach, or walk 1.3 miles north to Jeremy Inlet, the outlet of Scott Creek, a noted must-see spot to find shells and shark teeth. The more adventuresome shell seeker might arrange for a visit to Shell Island, a nearby sandbar where sand dollars and horseshoe crab shells can be found. Timing is critical to beachcombing success, and the very best time to inspect Edisto Beach is during an outgoing tide, during a new or full moon, or after a storm has passed through the area.

The true paleontological core of the Edisto fauna lies 60 to 100 miles offshore, where fossils lie embedded in submerged rock deposited during the late Pleistocene, when sea level was 150 feet lower than today. The ice age assemblage contains components of the Rancholabrean fauna, a varied collection of some one hundred vertebrate species of large-bodied animals, many of which were herbivore grazers. These vertebrates migrated to the New World by way of the Bering Land Bridge, the turnpike that connected Asia with North America during the ice age.

An amazing array of bone fragments and teeth of long-extinct terrestrial and marine animals, commonly turned black by the passage of time, wash onto Edisto Beach from the offshore strata. The cavalcade of beasts includes mammoths, mastodons, ground sloths, giant armadillos, giant beavers, long-nosed peccaries, and ancestral horses that died out long before Spaniards introduced new stocks early in the sixteenth century. Marine specimens include globular ear bones of whales, dolphin vertebrae, curved rib remnants of sea cows, and the piecemeal remains of sea turtles.

Fossil-hunting success at Edisto Beach is comparable to digging into a box of that well-known brand of molasses-flavored popcorn and peanuts. Inside is a mystery toy surprise, comparable to true Edisto fauna fossils, embedded in several handfuls of candied popcorn, analogous to the numerous shells found along the beach. The first handful from the box may result in a lot of sticky corn, but eventually you'll find a prize.

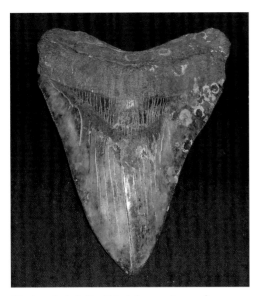

Shark teeth, including 7-inch-long specimens of Carcharodon megalodon—the largest shark that ever lived—can be found at Edisto Beach. Generally, however, only fragments of this shark's teeth are present. —Courtesy of the South Carolina Department of Parks, Recreation, and Tourism

Because Edisto Beach is located within the highest sea-energy sector of the eastern coastline, storms often leave in their wake a host of contemporary fossils, such as this palm-sized conch shell. —Courtesy of Rachael Garrity

Edisto Beach fossil detritus includes blackened fragments of mammoths, mastodons, bison, and horses that lived during the last stage of the Pleistocene Epoch, 50,000 to some 11,000 years ago. Pen for scale.

77. Mammoth Site, South Dakota

43° 25′ 29″ North, 103° 28′ 58″ West

The largest concentration of Columbian and woolly mammoths ever found.

Nestled among the rolling hills of southwestern South Dakota and named for its thermal waters, Hot Springs is home to Mammoth Site, an entry point into the Pleistocene world of frost and ice. The origin of the word *mammoth* dates back to early 1600s Europe and a report that made reference to "maimanto" tusks discovered melting out of Siberian tundra. Gargantuan in size and elephant-like in appearance, mammoths evolved in Africa around 50 million years ago, migrated to Europe and Asia, and crossed the Bering Land Bridge that at one time connected Siberia to North America. At the height of their reign in the New World, they grazed and browsed as far south as Central America. Their near-extinction 11,000 years ago is still a matter of debate, with either climate change or overhunting by the Paleo-Indian Clovis culture the more probable causes. Evidence shows that remnant herds populated a small island off Alaska as recently as 5,800 years ago.

Two members of the proboscidean family—characterized by tusks and a long, muscular trunk—proliferated on the American prairie: the Columbian mammoth and its smaller cousin, the woolly mammoth. The Columbian was a lumbering brute that stood 14 feet tall—so large the contemporary African elephant could walk under its chin—weighed in at 8 to 10 tons, and spent the better part of its time finding and consuming the 700 pounds of nutrients per day needed to satisfy its appetite. Small in comparison, the woolly topped out at 10 feet and 6 tons. Both were vegetarians; analyses of fossilized dung indicate they preferred grass, sedges, and a variety of marsh plants.

Researchers have employed a tusk count of two per animal to identify the remains of fifty-eight Columbia and three woolly mammoths at Mammoth Site. Pelvic bone analyses show the majority of them were juvenile males, expelled, perhaps, from a matriarchal population of adults. One skeleton, dubbed "Napoleon," is intact bone-wise, but the headless "Murray" may be the star attraction of the exhibit. Originally named "Marie Antoinette," the moniker was quickly changed when it was discovered that she was a he. Mammoth Site remains are not fossils in the normal sense of having been *permineralized*, a process whereby mineral matter replaces the original organic components. Instead, they are composed of original fragile bone and have therefore been left *in situ*, meaning they remain at the very spot at which they died.

Hot Springs is built on a bedrock foundation of Permian-age Minnelusa Formation limestone that is very susceptible to dissolution by warm, percolating groundwater. Approximately 26,000 years ago the roof of a cavern that had developed in this rock collapsed, creating a death pit that slowly filled with sediment during the following seven centuries, but not before countless mammoths and other ice age mammals were attracted to its artesian spring. For the unwary, the curious, and those inclined to risk-taking behavior, a site valued for life-sustaining water became a slippery slope, at the bottom of which animals became entrapped in gooey mud.

Located at 1800 US 18 Bypass, the 174-by-98-foot, 65-foot-deep, sand-and-clay-filled sinkhole is housed within a state-of-the-art working museum and laboratory, where excavation continues on a seasonal schedule, exposing a lost-world scene in which the arctic woolly became entombed with the temperate Columbian. This deathbed association of cousins—the only place both species have been found together—more than justifies the curator's claim that Mammoth Site is not merely a great museum, but indeed a mammoth mausoleum.

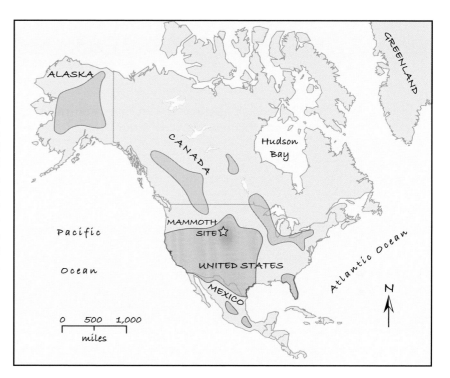

Generally, Columbian mammoth fossils are found in the United States, Mexico, and Central America (orange), while woolly mammoth fossils are principally found in the Great Lakes region north to Alaska (green). Both types exist at Mammoth Site.

Close-up of a mammoth skull and tusks at the Mammoth Site. Typically, male Columbian tusks were 10 to 14 feet long, not much longer than those of woolly mammoths. Female tusks were smaller and thinner.

78. Petrified Wood Park, South Dakota

45° 56′ 23″ North, 102° 09′ 33″ West

Whimsical, quirky, and eccentric—an artistic and unique exhibit that is both educational and photogenic.

It was most unusual for the elected officials of the small rural community of Lemmon to consider, let alone sanction, a parade comprising brass bands, drum and bugle corps, and an assembly of Native Americans from ten differing localities, many adorned in traditional war dress. After all, it was 1932, a national depression was in full swing, jobs were not to be found, and money was indeed scarce. And yet, it seemed the thing to do, especially since the star attraction would be the Sioux City Zeppelin, a 25-foot-long, wooden, linen-covered vehicle powered by a rear-mounted airplane propeller and coated in eye-catching silver paint. The jaunty zeppelin at the head of the parade was meant to provide pomp and ceremony for the dedication of Petrified Wood Park.

Millions of years ago—60 million to be exact—the region around Lemmon looked very much like coastal Mississippi does today: a subtropical scene of Everglades-like swamps and sluggish, meandering rivers. The Cannonball Sea, the last marine invasion to inundate North Dakota, was in full retreat. Sediments deposited in this sea—which became mudstones and sandstones of the Cannonball Formation—soon developed a soil profile favorable to a variety of luxuriant forests. Cypress, tamarack, sequoia, and hemlock grew in the lowland areas, while poplar, sycamore, elm, hickory, and hackberry took root in higher, drier elevations. Today, Paleocene-age petrified logs, stumps, and a plethora of broken and fragmented limbs provide ample evidence of these ancient forests.

Petrified wood can develop when wood is rapidly buried, thus preventing normal decay, and minerals replace the organic material. Common methods of burial include flood-stage rivers depositing layers of sediment, and volcanic eruptions or windstorms covering the forest floor with airborne material. Percolating mineralized groundwater seeps through the wood, coating its intricate cell walls and filling in the intercellular cavities. This transformation is often precise enough that details of bark, knots, and growth rings are maintained with remarkable clarity. Impurities in the groundwater add color to the fossilized wood: manganese contributes a purple hue, carbon forms black color, while iron creates earth tones of red, brown, and yellow.

Many yards, driveways, and gardens throughout northwestern South Dakota are embellished with specimens of petrified wood. Farmers reportedly concentrate this nuisance material into fencerows, much like their pioneering peers in the East created stone fences by clearing fields of glacial erratics. Ole Quammen, an amateur geologist and community visionary, decided in 1930 to utilize this abundance of fossil wood to increase employment and tourism. Some three dozen out-of-work citizens were hired to gather truckloads of permineralized wood within a 25-mile radius of Lemmon. The 300 tons of scavenged material—composed principally of Paleocene-age fossil wood, including tree trunks more than 30 feet tall, a secondary mixture of mammoth and dinosaur bones, and large and small spherical concretions—were then assembled into a city block–wide outdoor museum of eccentric sculptures. One particular sculpture is constructed of rock slabs tattooed with supposed dinosaur claw marks. Some paleontologists suggest these trace fossils may have been the result of male dinosaurs engaging in not-so-friendly competitive gyrations aimed at catching the eye of females during mating seasons.

This collage of petrified rock pyramids, arches, cones, and spires, distributed around a castle, wishing well, and waterfall, is both a monument to an ancient age and the world's largest petrified-wood peculiarity. The adjacent museum is completely constructed of petrified logs, and in its floor you can observe petrified slabs of grass. Located at 500 Main Avenue, Petrified Wood Park remains an off-the-main-drag curiosity second to none in uniqueness.

Several cone-shaped sculptures in Petrified Wood Park are constructed of spherical, compacted sandstone concretions that weather from rocks exposed in nearby river valleys. The largest of these "cannonballs" approach 6 feet in diameter.

This 300-ton multispired and gated grotto is constructed around an interior room in which a myriad of Cretaceous-age dinosaur bones are prominently displayed in the walls. —Courtesy of Thomas Dickas, Wausau, Wisconsin

A wall built of a hodgepodge of entire logs, broken limbs, and a variety of fossil wood fragments from a 60-million-year-old forest. —Courtesy of Thomas Dickas, Wausau, Wisconsin

79. Coon Creek Science Center, Tennessee

35° 20′ 04″ North, 88° 25′ 50″ West

Life in Tennessee in the Cretaceous Period was both productive and perilous.

At any one time during the Phanerozoic Eon, life-forms on Earth were preoccupied by bullies, transitory organisms that momentarily held the alpha position on the food chain. During the Ordovician Period it was *Isotelus maximus*, the up-to-1.5-foot-long trilobite that prowled the ocean floor with a ravenous appetite. Millions of years later the spotlight focused on *Dunkleosteus*, the 35-foot-long placoderm that delivered a bite force of 80,000 pounds per square inch and topped the scales at 4 tons. *Tyrannosaurus rex* is considered the master of the Cretaceous Period by many, but its reputation has been superseded by *Spinosaurus*, the crocodile-snouted, 50-foot-long, 10-ton carnivore that scavenged swamps and rivers worldwide 95 million years ago. More recently, the imposing saber-toothed cat of lore and legend, *Smilodon populator*, occupied center stage. Weighing in at up to 800 pounds, it cleansed the Pleistocene landscape of victims as varied as ground sloths and mammoths. Today, *Orcinus orca*, the robust, apex predator better known as the killer whale, plunders the world's oceans from the Arctic to the Antarctic at speeds up to 35 miles per hour, feasting with ease on a self-determined gourmet diet.

These notorious carnivores stain the pages of geologic history much like Jack the Ripper and Bloody Mary soil the pages of contemporary history. Additional bullies are recognized but are less well known. The fossil assemblage of the 75-million-year-old rocks that define the Coon Creek Valley of McNairy County, Tennessee, contains a prime example of a lesser-known ruffian—the mosasaur.

Sea levels were high during the Cretaceous Period, resulting in the formation of the Mississippi Embayment, a thumb-shaped advance of the ancestral Gulf of Mexico that extended as far north as southern Illinois. McNairy County was inundated by shallow, oxygenated, semitropical water teeming with marine crocodiles, sea turtles, crabs, lobsters, a variety of bivalves, sharks, and other fanged fish. An exemplary variety and number of fossil specimens of these organisms are found within the banks of Coon Creek, and the degree of preservation suggests the animals were entombed quickly after their death.

Mosasaurs, a class of bizarre marine reptiles that dominated the worldwide Cretaceous seas, were the reigning bullies of the Mississippi Embayment. The remains of two—identified as *Prognathodon overtoni*—were found embedded in the sandy marl of Coon Creek. Examination of their jaws and teeth suggests this denizen of the deep specialized in preying on ammonites, turtles, and a variety of other armored organisms using its robust bite force to first crush the protective shells and then shear the internal flesh. These 30-to-40-foot-long apex predators became extinct, along with the dinosaurs, 66 million years ago.

The Coon Creek Science Center (2985 Harding Graveyard Road) north of Adamsville, operated through the Pink Palace Family of Museums in Memphis, is open to the public by reservation only. Group-rate programs consist of both two- and four-hour excursions and include classroom discussion and field collecting and laboratory preparation of you-can-take-them-home fossils. Stephen Jay Gould, the best-selling science writer and evolutionary paleontologist, remarked that Coon Creek was "one of the twelve most important fossil sites in the United States" (Ehret, Harrell, and Ebersole 2016). With an inventory of animal and plant species approaching one thousand, Coon Creek has easily earned international recognition as a lagerstätte, a sedimentary deposit deemed remarkable for the quantity, diversity, and quality of preservation of its fossils.

Jointed legs and claw of a 75-million-year-old crab in the clay strata of the Coon Creek Formation. Dime for scale.

Specimens of Pterotrigonia thoracica, *the official state fossil of Tennessee, are found at the Coon Creek Science Center in museum-quality condition. Noted for its external ornamentation and affectionately known as "Ptero," this bivalve fed by burrowing through the mud floor of the Mississippi Embayment.*

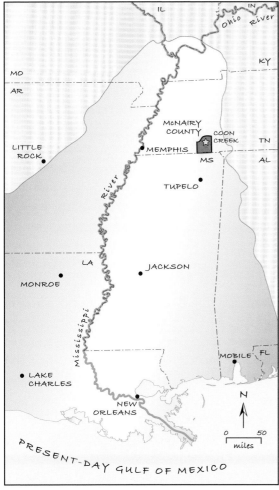

During Cretaceous time, the shoreline of the Mississippi Embayment (yellow), a high-water advancement of the Gulf of Mexico, was positioned as far north as southern Illinois. The star shows the location of the Coon Creek Science Center.

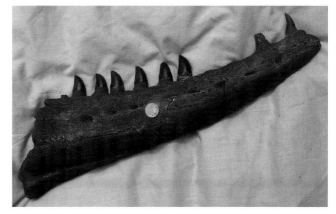

Lower jawbone of Prognathodon overtoni, *the predatory mosasaur found at Coon Creek. The jawbone contains eight of fourteen original teeth, extending an average of 1.3 inches from the fossilized bone. Penny for scale.*

80. Gray Fossil Site, Tennessee

36° 23′ 14″ North, 82° 29′ 54″ West

The world's largest cache of tapir fossils entombed in 4-to-7-million-year-old clay.

Periodically, the news reports that all too often provide less-than-welcome information also include items that focus on achievements in the worlds of science and technology, a respite from political bickering, crime reports, and other contemporary concerns. For example, the inquiring public can become enthusiastically involved in coverage of a new fossil find, calling upon evolutionary biologists, geologists, and paleontologists to fill in the details: Has a new genus been found? Is it extinct? Should it be labeled prey or predator?

The process sparks a series of press conferences and the assembly of multidisciplinary teams. Detailed analyses lead to the development of grant proposals and scholarly papers, and eventually a new chapter of geologic history is written. Once again, the catalogue of world knowledge expands, and humankind is enriched. This is precisely the remarkable story surrounding a discovery along Tennessee 75, west of Gray, in May 2000.

Because far too many automobile accidents had occurred along this stretch of road, the highway department decided to reroute some curves. When workers unearthed an unusual pocket of soft gray-to-black clay, investigation indicated the deposit was confined to a steep-sided depression in Ordovician-age dolomite. Closer study proved there were bone fragments ornamented with teeth strewn within the clay. Abruptly, road construction gave way to scientific inquiry: What caused the depression? Were the bones human or animal? How old were they, and was the site worthy of preservation and recognition?

Scientists considered four possibilities for the origin of the 300-foot-wide and 100-foot-deep channel: erosional, glacial, fluvial, and tectonic. The number was quickly cut in half; Tennessee had not been directly affected by glaciation during the ice age, nor had it experienced any significant degree of tectonic activity for millions of years. That left erosional and fluvial possibilities as culprits. In time, they determined the deposit was erosional in nature, more specifically karst, a dissolution topography that develops on and within carbonate bedrock and is characterized by sinkholes, caves, and underground drainage.

The agenda then moved to an inventory of the bones—their number, type, and genus classification. Many were the expected remains of frogs, mice, rabbits, and a variety of rodents, but others hinted at a possible history-altering discovery. A tusk and the 3-foot-long pelvic bone of a prehistoric elephant, along with the skull—a triangular-shaped mass indented with dollar-sized sockets—of the first fossil alligator ever found in Tennessee, gave the site paleontological pizzazz. Other bones were identified as those of a pot-bellied rhinoceros, humpless camel, short-faced bear, and collared peccary. A tooth fragment was suggestive of a saber-toothed cat, and two new prehistoric species were found: a relative of the extinct Eurasian badger and an extinct red panda. The biggest surprise, however, was the discovery of bucketloads of bones identified as those of tapirs, a short-legged, stubby-tailed, piglike mammal that disappeared from North America some 10,000 years ago. Representatives of all life stages—juvenile, adult, and senior—of this herbivore were found.

This bone-laden sinkhole may have functioned as both a death trap and a thirst-quenching habitat for a menagerie of inquisitive animals during Miocene time—just before the dawn of the ice age—but today it is the site of the multistory Gray Fossil Site and Museum (1212 Suncrest Drive) in Gray, home to an entire new chapter in Tennessee history.

Today fewer than two thousand red pandas roam the bamboo forests of Nepal and China, but 5 million years ago their ancestors, of the genus Pristinailurus, roamed bamboo forests in Tennessee, as evidenced by this red panda skull found at the Gray Fossil Site. Three-quarter-inch bar for scale. —Courtesy of the Center of Excellence in Paleontology at East Tennessee State University

Found in large numbers, tapir bones, including this outstanding skull and jaw, were the first fossils found at the Gray Fossil Site. Many specimens are fragmented, but no evidence of predation or scavenging has been found. —Courtesy of the Center of Excellence in Paleontology at East Tennessee State University

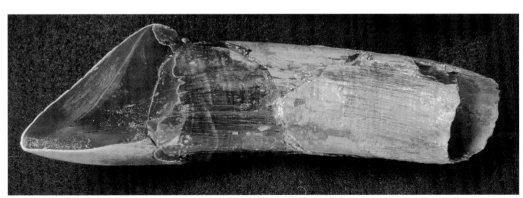

The tusk of an extinct Teleoceras, *a tubby grazing rhinoceros, unearthed at the Gray Fossil Site. Analyses suggest it was male and likely lived only half of its potential life span of ten to twenty years. Tusk is 18 centimeters long.* —Courtesy of the Center of Excellence in Paleontology at East Tennessee State University

81. Rock Island State Park, Tennessee

35° 48' 29" North, 85° 37' 60" West

Crinoids are perhaps the first fossils acknowledged as such by humankind.

Some fossil enthusiasts collect miscellaneous varieties of ancient life, some are interested in a specific time period, and others restrict themselves to a designated geographic area. Those whose passions focus on a particular class or phylum are niche enthusiasts. Rock Island State Park, noted for a variety of fossil body parts belonging to crinoids, a class within the echinoderm phylum, which is composed of such familiar organisms as starfish, sea urchins, sea cucumbers, and sand dollars, is a niche-enthusiast haven. Many paleontologists consider crinoids to be the very first fossils recognized as ancient life by humankind.

The name *crinoid* is built with the Greek words *krinon* and *oeides*, meaning "lily" and "like," respectively, a nod to their overall appearance. They have been recognized as fossils for more than 2,000 years and have been collected by diverse individuals, ranging from ancient metaphysicists to present-day picnickers. The structure of most crinoids, both living and extinct, consists of the *holdfast* that anchors the animal to the seafloor; a flexible, segmented stem; the *calyx*, a cup-shaped structure containing the vital organs of mouth, digestive system, and anus; and a series of feathery arms that gather food, such as algae and larvae, from passing oceanic currents. These segments are held together by fleshy matter that gives the organism flexibility while alive but easily decomposes upon death, causing the skeleton to break into fragments easily scattered by sea currents. Only rarely is the fossil animal found in a true-to-life form.

Stems and stem segments are by far the most common crinoid fossils, but because they occur throughout much of the fossil record and are not generally recognized as index fossils, they are the despair of many field paleontologists. Stems are made of a linear chain of disk-shaped segments known as columnals, which are generally circular or polygonal, and sometimes pentagonal, in plan view. A central canal passes through the length of the stem, and each columnal thus bears an axial perforation which also varies in design from circular to pentagonal.

Stems longer than 65 feet have been found, but more commonly they range between 1 and 3 inches. Columnals are commonly less than 0.5 inch in diameter, but some exceed 1 inch. In some Paleozoic-age carbonate rocks, columnals and stems constitute a major part of the rock, hence the name crinoidal limestone. Children, on the other hand, enamored by the range of shapes, often refer to the individual columnals as "Indian beads."

At one time considered extinct, crinoids were added to the list of living organisms when oceanographers began to explore the depths of the oceans. More than ten thousand specimens were dredged from the waters of Martha's Vineyard in a single haul in 1881, and twenty-five years later the deck of the *Albatross* was weighed down with tons of squirming crinoids on several dredging occasions during its expedition to the Bering Sea.

Rock Island State Park, north of Campaign, lies on a bedrock foundation of 340-to-330-million-year-old limestone deposited in a tropical, shallow, warm sea that inundated most of the North American continent during the Mississippian Period. Crinoid stems and columnals are present in riverbank exposures along both the Upstream and Downstream Trails, which depart from the scenic overlook at Twin Falls. These trails constitute a walk through deep time, when carbonate-producing sea lilies experienced a spurt of explosive evolution, known in the world of paleontology as the Age of Crinoids.

Twin Falls cascades over Mississippian-age limestone.

A mystery caused by differential erosion: Do the interior white segments represent crinoid stems, or are they the remains of central canals that have been filled with erosion-resistant material? Dime for scale.

Two segments of a broken crinoid stem. The upper fragment is constructed of at least seventy columnals. Penny for scale.

A columnal-coated slab of limestone. Note the broken stem immediately below the pen. The circular, central perforation is quite evident in many of the columnals. Pen for scale.

82. Fluvanna Roadcut, Texas

32° 57′ 28″ North, 101° 08′ 50″ West

Echinoids, devil's toenails, and steinkerns are an unexpected attraction at this rural exposure.

Having heard of a cluster of seven cities filled with gold and silver, Francisco Vázquez de Coronado formed an expedition of fellow conquistadors and headed north out of Mexico in search of wealth and fame. By the autumn of 1541, camped in eastern Kansas, he finally realized the cities of splendor did not exist. Before returning home, however, he composed a letter to the king of Spain describing a landscape he had crossed six months earlier: "I reached some plains so vast that I did not find their limit anywhere I went, although I travelled over them for more than 300 leagues" (Temple 2008). More than three centuries later, Captain Randolph Marcy, in charge of an expedition into the headwaters of the Canadian and Red Rivers, described the same land as "much elevated . . . almost total absence of water . . . even Indians do not venture to cross it" (Temple 2008).

Both Coronado and Marcy had experienced the physical isolation of the Llano Estacado, a 37,500-square-mile, flat, semiarid plateau covering a portion of the Texas Panhandle and an adjacent region of New Mexico. Crafted by the same tectonic forces that gave birth to the modern Rocky Mountains more than 66 million years ago, this dry and treeless plain is armored by a deposit of calcium carbonate–hardened soil that overlies flat-lying sedimentary rocks ranging from Permian to Cretaceous in age. Its eastern limit is marked by the Caprock Escarpment, a precipitous, 300-foot-high cliff incised by easterly flowing rivers and highways. One of these incisions, County Road 1269, transects an isolated portion of the Llano Estacado occupying the northwest corner of Scurry County, 5 miles north of the community of Fluvanna. The treasure envisioned by Coronado cannot be found here, but there is biologic "gold" for modern-day explorers interested in fossil treasure.

Three Cretaceous-age limestones are exposed within the Fluvanna roadcut; from oldest to youngest they are the Walnut Clay, Comanche Peak, and Edwards Formations. These strata contain a variety of fossils: echinoids, oysters, and a category of unusual trace fossils that gives the site special significance—steinkerns.

Echinoids have inhabited the seas since the dawn of the Ordovician Period, 485 million years ago, and are still present today, living encased in circular or heart-shaped calcareous shells generally less than 4 inches in diameter. Known also as sea urchins, they browse on algae and worms and extract organic debris from ocean-bottom sediment. Early Cretaceous specimens from West Texas are valued as index fossils useful for identifying the age of their host strata and the environment in which they lived.

Oyster fossils, including the extinct genus *Gryphaea*, abound at this roadcut. They lived on the Cretaceous seafloor, probably in large colonies, and consisted of two valves. The larger, gnarly one is distinguished by prominent growth lines that give it a supposed resemblance to Lucifer's cloven hoofs, which is why they are often called "devil's toenails." The tendency toward spiraling, however, may have contributed to their extinction. With too much curling in this valve, the counterpart flat valve could no longer effectively open or close.

Steinkerns, German for "stone kernels," are three-dimensional fossils of the consolidated sediment that filled the hollow interiors of fossil shells, which later dissolved or disintegrated. Many Lone Star fossil snails and clams are found almost exclusively as steinkerns. Those found at the Fluvanna roadcut, catalogued as snails and chalk white in color, are easily recognized by their coiled architecture.

Looking north along County Road 1269 and the Fluvanna roadcut, composed of three Cretaceous-age limestones: Edwards (top), Comanche Peak (middle), and Walnut Clay (along the base of the hill).

Fluvanna roadcut echinoids are easily recognized by their flattened, globular shell that exhibits radial, five-rayed symmetry. Penny for scale.

Before paleontology became a science, it was not uncommon for European countryfolk to carry a Gryphaea fossil in their pocket in the belief it would help prevent rheumatism. Penny for scale.

The original shells of these steinkerns weathered away, leaving internal molds composed of consolidated sediment. Note two of the molds underwent a degree of compaction, as seen by the shrinkage cracks. Penny for scale.

83. Ladonia Fossil Park, Texas

33° 27′ 25″ North, 95° 56′ 33″ West

A site harboring mosasaur, mollusk, coprolite, and miscellaneous tooth fossils.

Historic and archaeological records show that the first wave of Old World settlers arrived in northeast Texas in the late 1830s and built homesteads on land that Native Americans had populated more than 10,000 years earlier. By the 1920s the ancestral inhabitants were long gone, but the swampy conditions along the flood-prone, winding course of the North Sulphur River had become a vexing problem to the pioneering population. The solution seemed simple: change the meandering river channel to alleviate the flooding and create new farmland.

Modifying the river proved doable but problematic. Excavation in the 1930s created a straight, 20-foot-wide, 10-foot-deep channel that solved the flooding problem by improving the drainage. This alteration, however, greatly increased the rate of water flow—when meanders are removed a stream drops the same elevation but over a shorter distance, and the slope and velocity are both increased—an erosion process that over time widened the riverbed to 250 feet and deepened it to some 60 feet. Today the banks are eroding at an unacceptable rate of 1 foot per year.

Plans are now in the works to dam a lake, a proposal that would simultaneously solve an eighty-year-old environmental problem and create a defense against any future water supply crisis. One dilemma, however, remains: erosion continues to expand the channel, thus exposing fossils, and this stretch of river has become one of the nation's most visited areas for enthusiasts seeking Cretaceous-age mosasaur and plesiosaur remains, as well as an assortment of vertebrate and invertebrate fossils. Ladonia Fossil Park, situated along Texas 34, 2.2 miles north of the community of Ladonia, is the heart of this fossil-rich bottomland. A dammed lake would inundate this treasure trove of extinct life.

Between 80 and 70 million years ago, when *Tyrannosaurus* and *Triceratops* were beginning their land-based reign of terror in search of food, mates, and territory, their offshore vertebrate cousins—equally voracious and domineering—prowled the seas with similar intent. A relative to present-day lizards and snakes, mosasaurs were crocodile-like masses of doom and death that swam by undulating their 50-foot-long, barrel-shaped body. Prey that escaped their conical, razor-sharp teeth were likely to be consumed by smaller plesiosaurs eager to feast on leftovers. Analyses of mosasaur bite marks and coprolites suggest these alpha predators fed on a variety of victims, including turtles, large fish, squids, ammonites, plesiosaurs, and even other mosasaurs.

An abundant and diverse catalogue of Cretaceous invertebrate fossils is also found in the clay and mudstone strata forming the riverbed of the North Sulphur River. The saucer-sized oyster *Exogyra* is especially common and easily identified in gravel bars at arm's length. More up-close-and-personal methods of exploration are needed to find less-common collectibles, such as gastropods, pelecypods, echinoids, and ammonites.

Another appeal of Ladonia Fossil Park lies in the fact that while the riverbed strata contain fossils of Cretaceous-age marine life, the easily eroded rocks that form the adjacent riverbanks harbor land-based fossils that range in age from 12,500 years ago to modern times. Fragments of mammoth and mastodon tusks and molars are commonly found, and fossil evidence of horses, bison, llamas, ground sloths, camels, and woodland musk ox have also been collected.

When and if the proposed lake is completed, plans have been made to develop a new, nearby fossil-collection location reported to be even better than the Ladonia locale. Until then, the Ladonia site is the place to be on a balmy afternoon, especially after a heavy rain has unearthed new material and the current has returned to a shallow, non-threatening flow.

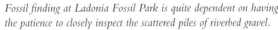
Fossil finding at Ladonia Fossil Park is quite dependent on having the patience to closely inspect the scattered piles of riverbed gravel.

Numerous fossil odds and ends are found at the Ladonia site, but many enthusiasts are not satisfied until they discover coprolites, fossilized shark excrement. The coprolite at left is 1.6 inches long.
—Courtesy of PK Kirkpatrick, Ladonia, Texas

Fossil teeth that can be found at the Ladonia site include the fang (this one is broken) of the extinct, predatory bony fish Enchodus, *also known as the saber-toothed herring (top row, center); specimens from* Tylosaurus *(two in the center), a marine mammal comparable to the present-day killer whale; and shark teeth (three in the lower left); among many others. The* Tylosaurus *tooth in very center is 1.75 inches long.* —Courtesy of PK Kirkpatrick, Ladonia, Texas

A 6-inch-long segment of a mosasaur jawbone showing three tooth sockets.
—Courtesy of PK Kirkpatrick, Ladonia, Texas

Gastropods from Ladonia Fossil Park are normally preserved as internal, phosphate-rich molds. The 2-inch-long specimen (left) is the shallow-water Turritella trilira, *whereas the smaller one is* Gyrodes major. —Courtesy of PK Kirkpatrick, Ladonia, Texas

84. Waco Mammoth National Monument, Texas

31° 36′ 25″ North, 97° 10′ 32″ West

Ice age evidence suggesting that mammals have always had to deal with slippery slopes.

In all probability, we will never know with certainty the events that took place 65,000 years ago near the confluence of the ancestral Bosque and Brazos Rivers in southeastern Texas. Humans were seemingly not involved, and it appears that the fossil remains weren't disturbed by scavengers. Guided by detailed soil investigations and age dating, however, many experts agree that flash flooding was probably responsible for the largest known concentration of ice age mammals to have died during the same event.

Two youths discovered the fossil site, now part of Waco Mammoth National Monument (6220 Steinbeck Bend Drive), in the spring of 1978 while searching for arrowheads on the outskirts of Waco. At first the specimen they noticed did not seem large, but when it was fully exposed it measured more than 4 feet long and a full 20 inches around. Experts at nearby Baylor University identified it as the femur of a *Mammuthus columbi*, otherwise known as a Columbian mammoth. Organized excavations were soon undertaken, and over the course of the next two decades the remains of twenty-four Columbian mammoths were uncovered, an entire herd, along with fossils of other Pleistocene Epoch vertebrates, including those of the saber-toothed cat, American alligator, giant tortoise, dwarf antelope, and western camel.

Mammoths were large, elephant-like mammals that evolved in Africa, migrated to Europe and Asia 2.5 million years ago, and later crossed into North America by way of the Bering Land Bridge. At the height of their reign, they grazed and browsed over a territory that extended from the latitude of Nebraska—the approximate southern limit of permafrost during the Pleistocene Epoch—to the grasslands of Central America. They became extinct around 11,000 years ago, either hunted out of existence by early humans or killed by habitat change that occurred when ice age glaciers were melting and Earth was entering a period of global warming.

Weighing in at as much as 10 tons, an adult Columbian male stood 14 feet tall—so large the contemporary African elephant could walk under its chin—and spent most of its day satisfying its nutritional needs, consuming 500 to 700 pounds of vegetation and drinking 40 gallons of water. Their spiraled tusks, extensions of incisor teeth, grew up to 14 feet long and were used to forage for food and fight off competitors. Analyses of tusk growth rings show that Columbian mammoths had a life span similar to that of a modern-day human—some eighty years.

Excavations prior to 1998 uncovered a nursery herd of females and offspring believed to have died in a catastrophic flash flood while assembled in a steep-sided river channel. Bone distribution patterns suggest the adults were attempting, at their very moment of death, to simultaneously form a circular defensive position and use their tusks to move the juveniles to safety. These fossils are preserved at Baylor University.

Discoveries unearthed since 1998 at Waco Mammoth National Monument can be seen in the climate-controlled Dig Shelter. This exhibit focuses on the in situ remains of a large male mammoth with a healed rib—broken perhaps during a violent encounter with a rival male—a female, two juveniles, and a western camel. Tours are conducted year-round, Tuesday through Saturday. This fossil site presents a snapshot in time, when disaster overwhelmed a group of lumbering giants, subjecting them to a situation in which they either drowned or were buried in mud, a scenario from which they could not, and did not, escape.

Constructed over the actual site of excavation, the suspended walkway inside the Dig Shelter facilitates an up-close-and-intimate examination of a catastrophic event that exterminated an entire herd of unsuspecting ice age behemoths.
—Courtesy of the Waco Mammoth National Monument

Columbian mammoth tusks grew at an annual rate of 1 to 6 inches, gradually curving until the tips pointed toward each other. This partially exhumed 10-to-11-foot-long pair was the pride of an adult bull. —Courtesy of the Waco Mammoth National Monument

85. Mineral Wells Fossil Park, Texas

32° 49′ 33″ North, 98° 11′ 26″ West

A crinoid heaven, for both the critters and the fossil hunters.

Tiny may not be the proper word to describe their size; *minute* is a better choice, but *miniscule* is perhaps best. They have been variously known as star stones, Indian beads, St. Cuthbert's beads, fairy money, and St. Boniface's pennies and were among the first fossils noticed by humankind. Generations of field-oriented collectors have puzzled over their existence. Technically called *columnals*—the disarticulated remnants of the stems of crinoids—they are the most common fossils found at Mineral Wells Fossil Park (2375 Indian Creek Road), in Mineral Wells.

Living crinoids have been found worldwide in oceanic latitudes that extend from 50 degrees south to 80 degrees north. Commonly referenced as "sea lilies" because they resemble flowers, they belong to the same class of marine organisms as sea urchins, sea stars, and sand dollars. Crinoids first appeared 480 million years ago, during the Early Ordovician Period. Expanding rapidly in diversity and number throughout the Paleozoic Era, most—but not all—died during the catastrophic episode of extinction at the end of the Permian Period, when as much as 96 percent of all oceanic species became extinct. The causes of this event are still the subject of heated debate. Prior to their near extinction, crinoids thrived in relatively clear-water "gardens" composed of immense populations of reef-forming creatures.

Complete crinoid specimens consist of several distinctive parts: the bud-shaped cup, or *calyx*, enveloped a mass of vital organs; arms, an undulating cluster of five appendages or multiples of five, captured and funneled food to the calyx; and a linear series of disk-shaped columnals formed the *stem*, terminating in the form of a *holdfast* (root system) that anchored the animal to the ocean floor. Calyx and arm fossils are rarely found at Mineral Wells Fossil Park, but columnals, and to a secondary extent stems of columnals, abound. Discovery success, however, is dependent on a willingness to engage in prone investigation, up close and personal with the rock debris, as the typical columnal rarely exceeds 0.3 inch in diameter, and the majority of stems are less than 1 inch long.

Mineral Wells columnals are mostly circular or, uncommonly, ovate. Punctured by a central canal that varies in outline, from circular to pentagonal, specimens often bear circular scars that mark the attachment points of fingerlike appendages that radiated outward from the stem of the living animal, providing additional anchorage.

While searching for columnals and stems, sharp-eyed enthusiasts have little trouble assembling a collection of beautifully formed *Chonetina* fossils. Semicircular in outline and less than 0.5 inch long, these brachiopod valves are easily identified, under magnification, by a series of stumpy spines projecting from the straight hinge. Bilaterally symmetrical, brachiopods are bottom-dwelling organisms that feed on bits of organic material. After populating the oceans in great numbers during Pennsylvanian and Permian time, *Chonetina* were exterminated by the Permian extinction event.

Once an old borrow pit locale, an excavated area dug to provide construction-grade sand and gravel, the Mineral Wells site was closed in the 1990s and then reorganized as a fossil park and opened to the public daily, sunrise to sunset. Twenty years of erosion of the Pennsylvanian-age Salesville Shale Member of the Mineral Wells Formation has exposed, in addition to crinoids and brachiopods, a variety of corals, pelecypods, bryozoans, shark teeth, and an occasional trilobite. However, this easily accessed and family-friendly site is a crinoid heaven of exceptional wealth. Look for the fossil identification sign located at the park entrance.

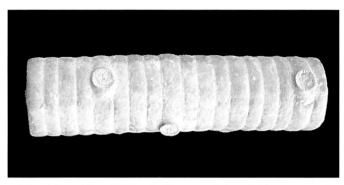

A 2-inch-long Mineral Wells crinoid stem consisting of eighteen columnals. Note the scars of three fingerlike appendages that radiated outward from the stem of the living animal, providing secondary anchorage. —Courtesy of Lance Hall

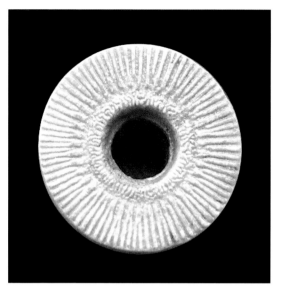

This 0.3-inch-diameter crinoid columnal has a circular, central canal that, in the living animal, housed bundles of nerves. —Courtesy of Lance Hall

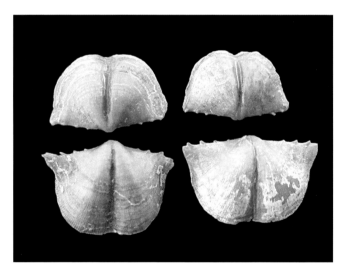

A myriad of 0.3-to-0.4-inch-wide Chonetina *brachiopods are available at Mineral Wells Fossil Park. Note the distinguishing stumpy spines attached to the straight hinges.* —Courtesy of Lance Hall

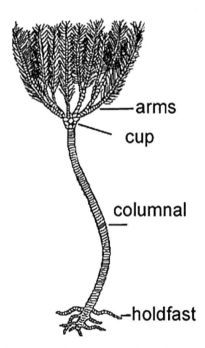

The anatomy of a typical crinoid. Heights varied, but one study of Ordovician-age species demonstrated that the stem lengths, from cup to holdfast, ranged from 0.5 to 12.5 inches. —Courtesy of Lance Hall

86. U-Dig Fossils Quarry, Utah

39° 21′ 18″ North, 113° 16′ 44″ West

This mecca for the trilobite fancier is home to both common species and unusual "bilobites."

Considered the signature organism of the Paleozoic Era, the trilobite is the coveted cornerstone of many fossil collections and the central attraction of many a weekend field trip. Evolving approximately 520 million years ago, these crab-like organisms peaked in biomass and diversity in Late Cambrian time; began a slow decline punctuated by brief bursts of adaptive radiation in Ordovician, Silurian, and Devonian time; and then became extinct 252 million years ago. They average 2 to 4 inches in length, but exoskeletons as small as 0.1 inch and as large as 27 inches have been found. Because their remains are always found with distinctive saltwater organisms, they are categorized as exclusively marine. It is believed they lived as both vagrant bottom scavengers and planktonic filter feeders.

The name *trilobite*—meaning "three lobes"—has long been a source of confusion. Some references emphasize the longitudinal segments that compose the anatomy of the exoskeleton: right and left pleural lobes and a central axial lobe. Others reference the fact that the body is divided laterally into three sections: cephalon (head), thorax (midsection), and pygidium (tail). Both appear to be adequate explanations for the animal's name.

Trilobites are found in many parts of the United States, but exposures of the 500-million-year-old Wheeler Shale in Utah are famous for Cambrian-age specimens. Arguably the best of all Wheeler Shale locations is the U-Dig Fossils quarry—advertised as the richest deposit of trilobites in the world—located 32 miles west of Delta via US 6/50. Turn right at mile marker 56.5 and follow the signs an additional 20 miles to the quarry. The fossils are found in calcareous shale that splits easily into flat sheets, revealing a collection of specimens that lie flat along bedding planes. Deposited during a relatively quiet episode of geologic history in a warm, shallow sea endowed with a nutrient-rich clay that supported a highly diverse fauna, the Wheeler Shale hosts more than forty species of trilobites. Six are generally found at U-Dig Fossils, with three being common: *Elrathia kingii*, *Asaphiscus wheeleri*, and *Peronopsis interstricta (Itagnostus interstrictus)*.

Possibly the most well-known trilobite on Earth, *Elrathia kingii* is found in such quantities throughout the United States that one commercial operator reportedly collected 1.5 million specimens in a twenty-year career. Averaging 1 inch in length, with a thorax usually constructed of thirteen segments and a cephalon accentuated by short spines, *Elrathia* scavenged the ocean floor in search of organic particles. Most *Elrathia* fossils lack cheeks, portions of the cephalon that broke away during—and thus facilitating—the molting process; the missing cheeks serve as proof that the specimen is a molt, not the fossilized remains of the deceased animal.

Classed as a bottom feeder that walked along the seafloor, *Asaphiscus wheeleri* is found in both North America and Australia, most commonly as a cheekless molt that averages 2 inches in length. It is characterized by an exoskeleton one-and-a-half times longer than wide, a thorax comprising seven to eleven segments, and a pygidium accentuated with a prominent wide, flat border.

Found in Asia, Australia, Europe, and North America, fossils of *Peronopsis interstricta* are easy to identify: they are generally less than 0.5 inch long, lack eyes, have a cephalon and pygidium that are nearly equal in size, and have a two-segment thorax. Following a hasty examination that overlooks the minimally developed thorax, one might believe he or she had discovered a "bilobite." Whatever the terminology employed, the frequency of trilobites found in certain layers of the Wheeler Shale justifies its classification as a lagerstätte that no trilobite enthusiast should overlook.

Getting down—and not so dirty—with hammer, chisel, and bucket is the preferred and recommended method of unearthing Cambrian-age trilobites long entombed within slabs of Wheeler Shale. —Courtesy of Robin Callahan

This impression of Elrathia kingii *is anatomically correct: it has a cephalon complete with a pair of cheeks; a thirteen-to-fourteen-segment thorax; and a small, unimpressive pygidium, suggesting it is the remains of a deceased specimen and not a molt. Penny for scale.*

With eight segments composing the thorax and a well-defined and flat pygidium, this fossil is easily identified as Asaphiscus wheeleri. *Note that both cheeks are missing, meaning this specimen is a molt. Penny for scale.* —Courtesy of Rachael Garrity

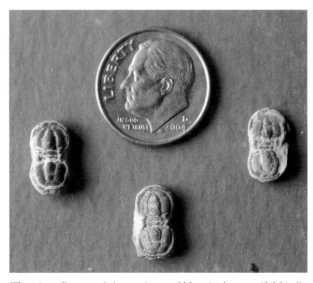

The minute Peronopsis interstricta *could be mistaken as a "bilobite," a reference to the fact that this blind trilobite appears to have only a cephalon and pygidium. Look closely, however, and you can see a two-segment thorax. Dime for scale.*

87. Wall of Bones, Dinosaur National Monument, Utah

40° 26′ 17″ North, 109° 18′ 29″ West

**Left intact as found, a 50-yard-long sandstone wall
is a window into the real world of *Jurassic Park*.**

Many mysteries associated with the history of Earth have been unraveled since the birth of the science of geology, but many remain. One of the more inconsequential of the unsolved category, at least to the average person, is how many geologic formations are there in the world. A *formation* is a body of igneous, sedimentary, or metamorphic rock with recognizable boundaries that can be traced in the field without detailed analyses and that is large enough to be represented on a geologic map. It is widely recognized as the basic cartographic unit used in field mapping. William "Strata" Smith, the father of English geology, referenced thirty-four Great Britain–based, hand-colored formations in his famous 1815 *Geological Map of England and Wales*, the first map to show the geology of an entire nation. Since that time, the list of formations recognized worldwide has grown to number in the thousands.

Despite their large number and general lack of crowd appeal, a few formations have gained international superstar status: the Solnhofen Limestone of Germany, home to the birdlike dinosaur fossil *Archaeopteryx*; an unnamed (to protect it) outcrop in the Jack Hills of Western Australia that hosts 4.4-billion-year-old zircons, the oldest-known material of terrestrial origin; and the Burgess Shale of Canada, famous for its exceptional collection of soft-bodied Precambrian fossils. In the United States, many paleontologists consider the Morrison Formation and the phrase "terrible lizards" to be synonymous.

Found throughout a wide swath of western states, the Morrison Formation is famous for the variety of dinosaur fossils it contains, including *Diplodocus*, *Apatosaurus*, *Stegosaurus*, *Allosaurus*, and *Camarasaurus*. Associated with a wide variety of Late Jurassic environments, the Morrison strata most likely to contain dinosaur fossils are those deposited within a river environment. Perhaps the most famous of these fluvial deposits are exposed along the south flank of the Uinta Mountains in Utah. Protected by Dinosaur National Monument, these rocks are the world-famous home of the "wall of bones." This 150-million-year-old hogback ridge of sandstone, which dips 67 degrees, has been laboriously quarried to reveal the remains of approximately 1,500 dinosaur bones, fossils of carcasses carried downstream until they became embedded in a sandbar, a ridge of sand deposited in or along the edge of a river.

Measuring 150 feet in length and several stories high, the wall was discovered in 1909 when eight *Apatosaurus* tail vertebrae were seen protruding from the ground in a lifelike position. The bone bed was excavated over the years and finally given protection within the Quarry Exhibit Hall, an all-weather enclosure. Eventually the fossil remains of approximately five hundred dinosaurs and other animals were discovered, including those of five sauropods (long-necked, long-tailed herbivores), three theropods (short bipedal carnivores), and three ornithischians (herbivorous dinosaurs with a birdlike pelvis). Millions of years of geologic history are exposed along the Fossil Discovery Trail, beginning at the trail junction adjacent to the Quarry Exhibit Hall. A trail spur hugs a sandstone cliff composed of river-deposited sand and gravel that contain fossil fragments of dinosaurs along with impressions of freshwater clams.

In 1993, millions of Americans viewed the science-fiction thriller *Jurassic Park*, featuring a fanciful lineup of terrorizing goliaths—*Dilophosaurus*, *Brachiosaurus*, and an ahead-of-its-time *Tyrannosaurus* (which didn't come on the scene until the Cretaceous Period). It quickly became the highest-grossing film of its time. If seeing the real life of the "Jurassic Park" past is on your to-do list, the wall of bones, though devoid of blood-curdling death-defying action, is the perfect contrast to Hollywood's make-believe.

As seen to the west from the Quarry Exhibit Hall wall of bones outcrop, Jurassic-age sandstone and siltstone dip steeply to the north. Rocks of similar age can be examined in detail along the nearby 1.2-mile-long Fossil Discovery Trail.

This 8-inch-long left shinbone, one of several fossils in the wall of bones that can be touched and felt, once was part of a Diplodocus, a 10-to-16-ton, long-necked, long-tailed herbivore that roamed the Utah countryside 150 million years ago.

Discovered in 1924 at Dinosaur National Monument, this 34-inch-long Allosaurus skull is one of the best preserved ever found. Allosaurus was a massive, short-necked predator that reached adult size by age fifteen, grew to a length of approximately 40 feet, and weighed more than 3,250 pounds.

88. Chazy Fossil Reef National Natural Landmark, Vermont

44° 51' 10" North, 73° 20' 24" West

The first-ever reef.

Many of the nature-loving tourists who fly into Harare, the capital city of the landlocked African Republic of Zimbabwe, located at 18 degrees south latitude, have but one thing on their mind: securing back-home bragging rights by taking photographs of the mammal world's "big five": buffalo, elephant, leopard, lion, and rhinoceros. If it were possible for the same visitors to be transported back 480 million years, to when the breakup of the supercontinent Rodinia was redefining world geography, photographic opportunities would be amazingly different. The tourists would still be at 18 degrees south latitude but would be standing along the austere, treeless shoreline of Laurentia, precursor to North America created by the death of the supercontinent Rodinia. Inland, liverworts and ancestors of today's mosses would be extending roots into damp and rock-shaded areas. To the east, members of a diverse cast of invertebrate organisms—trilobites, crinoids, and brachiopods—would be probing the depths of the Iapetus Ocean, precursor to the Atlantic. The focus of any snapped photo, however, would surely be Chazy Reef, a shallow-water structure that was the first known biologically diverse reef in the history of life.

Exposures of Chazy Reef, a marine structure that defined the southeast border of Laurentia in Early Ordovician time, have been traced from Newfoundland to Tennessee. One thousand miles long and 480 million years old, this feature fits the definition of a *reef*, a ridgelike structure composed of the calcareous remains of sedentary organisms such as corals, algae, mollusks, bryozoans, gastropods, and foraminifers. Glacially polished and publicly accessible remnants occur on Isle La Motte, the northernmost island of Lake Champlain, at the Fisk Quarry Preserve (4039 West Shore Road) and Goodsell Ridge Preserve (69 Pine Street). These locales are collectively recognized as the Chazy Fossil Reef National Natural Landmark.

Throughout Precambrian time cyanobacteria, the earliest fossil examples of life, built stromatolites, microbial mounds with a layered internal structure. These cabbage-like configurations dominated the shallow water of the world's oceans, but they did not form reefs. The Cambrian explosion introduced a variety of organisms, but none were colonial in nature, so they did not form reefs either. Then, with the dawn of Ordovician time, evolutionary variation produced colonial communities, and Chazy Reef was born.

The oldest layers of the reef are confined to the basement rocks of the Fisk Quarry, consisting of a limestone matrix that binds scatterings of tufted mounds that bryozoans constructed. These small, mainly seafloor-dwelling animals lived in colonies on the floor of the Iapetus Ocean. Younger, overlying layers contain stromatoporoids, sponge-related organisms that grew by secreting laminated, calcareous skeletons. Several-foot-high, convex-upward white domes in the Fisk Quarry walls are very visible examples of these creatures. The diverse, youngest layer of Chazy Reef, for which it is famous, is exposed at Goodsell Ridge Preserve. Bryozoans, stromatoporoids, and their look-alike cousin stromatolites, supported by accumulations of gastropods, cephalopods, algae, crinoids, and the occasional coral, compose this portion of the reef's architecture.

This sequence of old-to-young layers represents an increasing variety of ecological niches that sheltered an expansive inventory of marine animals. It is the first-ever example of the type of faunal succession and community life exemplified today by the Great Barrier Reef of Australia. Paleontologists are divided on what words to use to describe Chazy Reef. Some say it is the world's oldest coral reef, whereas others insist it is the world's oldest reef in which corals first appeared. Perhaps it is best to say the trendsetting Chazy Reef is the world's very first real reef community.

An uncoiled cephalopod shell exposed along the walkway at Goodsell Ridge Preserve. At least seven chambers can be seen in this glacially eroded specimen. Pen for scale.

A cross section of a pair of sponge-related stromatoporoids (white mounds)—the principle reef builders during Ordovician time—exposed in the walls of the Fisk Quarry.

A 480-million-year-old coiled cephalopod, ancient ancestor to today's octopus and squid, on view at the Museum and Learning Center at Goodsell Ridge Preserve. Specimen is 11 inches long.

89. Chippokes Plantation State Park, Virginia

37° 08′ 45″ North, 76° 44′ 21″ West

A mollusk that stands tall among all others.

The history associated with the discovery, classification, and naming of any one fossil can be both interesting and confusing. *Chesapecten jeffersonius*, a mollusk with a history dating back to the earliest years of the American Republic, is an excellent example. In 1687 Martin Lister, an English-born medical doctor and conchologist, published *Historiae Conchyliorum*, a pioneering text that included a drawing of *Chesapecten jeffersonius*, the very first North American fossil to be illustrated in a scientific publication. Lister graduated from St. John's College in Cambridge in 1658 and became a fellow of the Royal Society thirteen years later. Fascinated by many aspects of natural science, his publications are scientifically tainted by his belief that fossil mollusks were inorganic imitations of extant specimens found along the beaches of Europe and North America.

Visiting America in 1824, the English geologist John Finch assembled a large collection of mollusk fossils from the banks of the York River, in Virginia, and gave it to the Academy of Natural Sciences of Philadelphia. Fossils were so abundant in the York River area at this time that residents used them in the foundations of their buildings and as dishes and water ladles. Academy scientists were given the task of describing and naming the fossils, all of which were found to be unknown to science. One of the scientists, Thomas Say, cofounder of the academy and author of the six-volume work *American Conchology*, the first text published on this subject in America, named a particularly well-preserved specimen *Pecten jeffersonius* in honor of Thomas Jefferson, who many historians have sanctioned as the father of North American paleontology.

For decades it was thought that Finch had collected specimens of *Pecten jeffersonius* along the St. Marys River in Maryland, but research by Lauck Ward, a paleontologist with the US Geological Survey, proved that their provenance could not only be traced to Virginia, but that the fossils were associated with a younger geological formation than the one exposed in Maryland. Ward brought closure to this history of confusion by renaming the fossil *Chesapecten jeffersonius* in 1975, and since then it has become widely known to mollusk enthusiasts as Jefferson's Chesapeake scallop. In 1993 it became Virginia's official state fossil. It is found in abundance along the shore of the James River in Chippokes Plantation State Park, especially at low tide, when a larger beach and a less explored area is exposed.

Along with their oyster and clam cousins, scallops are bivalves consisting of two shells held together by a powerful muscle. Economically important in the prehistoric world, they have long been an important source of food and even served as currency. Native Americans used them for body adornment and as bowls and scraping tools. With a lineage that dates back to the early years of the Paleozoic Era, scallops occur in sedimentary rocks of many ages, have a wide geographic distribution, and are rather easy to identify. Bottom-dwelling types are sensitive to water depth and temperature and are therefore employed by the petroleum industry to reconstruct geologic environments that may harbor oil and natural gas.

Nine to twelve prominent radiating ribs and a rounded, undulating edge characterize *Chesapecten jeffersonius*. It spent its life in the shallow waters of the continental shelf from 10 to 4 million years ago, during late Miocene and early Pliocene time, before it became extinct, the victim of a cooler climate. During its youthful stage it lived attached to the seafloor, but many adults freely moved about, propelling themselves by rapidly opening and closing their shells.

After storm waves have stirred up shallow-water sediment, any number of odd marine body parts can be found at Chippokes Plantation State Park, including shark teeth, sturgeon scales, barnacles, whale bones, and sun-bleached sea turtle ribs, such as this specimen. Pen for scale.

Fossils of the extinct Ecphora quadricostata *are coveted by collectors because they are seldom found in an unbroken condition. This predatory sea snail bored holes through the shells of fellow mollusks and then fed on their vital organs. Specimens are about 1 inch across.*

Chesapecten jeffersonius.
—Courtesy of Rachael Garrity

90. Hogue Creek, Virginia

39° 14' 40" North, 78° 16' 43" West

Reconstructing deep-time landscapes and ages with fossils.

Fossils have value, but their specific worth depends on different points of view. To the novice they can be collectible objects, made into jewelry, used as paperweights, or treasured as a reminder of a memorable family vacation. In contrast, based on the knowledge that animals that frequent different environments are normally different in size, shape, and structure, paleontologists are able to use fossils to reconstruct environments that have occurred in the geologic past. An intact fossil encased in shale might represent an organism that lived along a muddy, low-energy seafloor, whereas broken shell segments embedded in coarse sandstone are suggestive of life in shallow-water, high-energy conditions.

Fossils are also the principle means for recognizing and identifying units of geologic time. William Smith, the father of English geology, recognized two hundred years ago that many rock units possessed distinctive fossil organisms. Known as *index fossils*, they existed across a wide range of longitude and latitude for a short period of geologic time. Trilobites are index fossils of the Paleozoic Era; their presence anywhere in the world is evidence that the host rock was deposited between 541 and 252 million years ago.

The field analysis of a small exposure of dark shale forming a hillside in rural Frederick County is very useful for demonstrating how scientists employ fossils to determine ancient environments and geologic age. Travel 2.5 miles north from downtown Winchester on US 522, turn west onto Virginia 679, and proceed 4.3 miles to a small, two-vehicle parking space close to the banks of Hogue Creek. Walk west 275 feet along the rail right-of-way to the outcrop.

The rock contains sparse, poorly preserved snail fossils and two categories of brachiopods: many pea-sized pygmy specimens too small to be effectively identified without a microscope, and an eye-catching distribution of shells with a distinctive morphology. Brachiopods are marine animals that have existed throughout the oceans of the world since the early days of the Cambrian Period. Also known also as lampshells, they were a dominant form of life that suffered a high, but not quite complete, degree of extinction at the end of the Permian Period. Some 350 species are known to live today, and more than 12,000 fossil species have been identified.

The attention-catching brachiopods at Hogue Creek belong to the extinct genus *Mucrospirifer*, otherwise called "butterfly shells" because of their distinctive, elongated winglike appearance. Studies indicate this genus of 1-to-1.5-inch-long animals preferentially lived in mud environments attached to the seafloor with a fleshy stalk and filtered dinner from the water that flowed around them. *Mucrospirifer* lived for only 39 million years—from 411 million years ago to 372 million years ago—an abbreviated period of geologic time that makes them index fossils of the Early and Middle Devonian Period.

With this knowledge the geologic lesson to be learned at Hogue Creek is simple: the presence of *Mucrospirifer* shells is evidence that during Early to Middle Devonian time the Winchester area of Virginia was inundated by a marine body of water characterized by a muddy seafloor environment. This analysis is supported by the fact that the host strata is shale, a sedimentary rock formed by the compaction of mud. Because *Mucrospirifer* is an index fossil, its presence in any area of the world is evidence of ancient conditions similar to those related to the Hogue Creek exposure.

The compacted dark-gray shale at Hogue Creek is part of the Mahantango Formation, which was deposited some 390 million years ago in a shallow, tropical sea covering portions of Pennsylvania, West Virginia, and Virginia.

Mucrospirifer *brachiopods can be identified by the deep fold down the center of the shell and their winglike appearance. Penny for scale.*

The Mucrospirifer *brachiopod is easily recognized, but until this exposure experiences additional weathering, the many bean-sized mounds dappling the surface cannot be effectively identified. Dime for scale.*

When the surface of the Mahantango Formation becomes thoroughly weathered, former nondescript mounds clearly become pygmy brachiopods. Penny for scale.

91. Mint Spring, Virginia

38° 04′ 36″ North, 79° 05′ 05″ West

Mystifying hacksaw-shaped critters that are either long-lived or long-extinct.

Anyone interested in arranging the myriad of invertebrate fossils found in the inventory of any major museum—for example, the Smithsonian Institution—into categories defined by evolutionary history might well be advised to use a tripartite system. One grouping would represent life that existed in the past and can still be found today, such as brachiopods, while another would contain life long gone, such as trilobites. The third group could have the most interesting creatures of all: those that some experts deem extinct, others extant. Many would agree that graptolites fall into this latter category.

When the eighteenth-century Swedish botanist Carl Linnaeus, father of modern taxonomy, first observed scraggly patterns on the surfaces of sedimentary rocks, he was befuddled. Gathering his thoughts, he announced that while they resembled fossils, they were actually the products of chemical processes. He named them *graptolites*, meaning "writings on rocks."

Today we know that graptolites are colonial animals that evolved during the Cambrian Period, became dominant in the Ordovician and Silurian seas, and then supposedly died out in Mississippian time. Colonies were constructed of a variable number of minuscule, eyelash-shaped branches edged by serrated structures, each a cuplike housing that sheltered a zooid, an individual member of the colony. Zooids strained plankton from the water with a cluster of undulating tentacles.

Historically, graptolites have been classified into two major orders: Dendroidea and Graptoloidea. The first to appear were Dendroidea, distinguished by a many-branched, bush-like body shape that resembled the tributary pattern of a large river system. They were bottom dwellers attached to the seafloor by a rootlike stalk. Dendroidea survived until the Mississippian Period. Graptoloidea were *pelagic*, meaning they drifted and floated freely in ancestral seas, hitchhiking on floating seaweed or held aloft by a gas-filled bladder. They differed from their dendroid cousins by having fewer branches, being more prolific, and reaching extinction sooner—in the early part of the Devonian Period.

Found on every continent except Antarctica, the total inventory of graptolite fossils exceeds two thousand species. Because they are common, have worldwide distribution, and evolved quickly into a multitude of species, they are highly valued as index fossils that identify and date rocks separated by great distances. The oil and gas industry extensively uses the presence of graptolites in formations; they serve as indicators of ancient water depths and temperatures, both of which can lead explorers to valuable fossil fuel reserves.

Graptolites are commonly preserved as a flattened, black carbon film on the surfaces of dark shale. These black-on-black specimens are, at first glance, often difficult to see. Occasionally specimens of a contrasting nature are found on light-colored strata, such as the fossils embedded in tan to cream-colored shale along Virginia 654, 100 feet east of the northbound ramp onto I-81 (exit 217), outside Mint Spring. The 1-inch-long sawtooth branches scattered across the outcrop, which dips at 85 degrees, may belong to either the *Nemograptus* or *Climatograptus* genus of graptolites, both of which were common in the Ordovician seas.

Several decades ago, the capture of curious creatures from the depths of the Pacific Ocean off New Caledonia caused considerable consternation in the world of paleontology. Today, their classification is still the subject of debate. Some investigators say they have the characteristics of ancient graptolites, while others reject such speculation as pure bunk. Until the scientific jury reaches consensus on what these modern creatures are, the geologic life span of graptolites will be discussed with vigor. Whether they are cast in the category of extinct or extant will depend largely on the most accepted opinion of the moment.

Graptolite fossils are commonly found as a collection of individual branches, or stipes, black carbonaceous films resembling scroll-saw blades. Each "blade," here about 1 inch long, is constructed of a series of overlapping tubes, each of which housed an individual microscopic organism.

Deep circular pits that mar the surface of the Lincolnshire Formation at Mint Spring are the result of black chert nodules weathering out of the Ordovician-age rock. Local rock hounds prize the nodules, which are composed of quartz and chalcedony.

The nearly slate-board smooth exposure of dipping rock at Mint Spring is one limb of a series of northeast-southwest-oriented anticlines that compose the Allegheny Mountains, a range of the Appalachian Mountains.

92. Ginkgo Petrified Forest State Park, Washington

46° 57′ 16″ North, 119° 59′ 17″ West

One of the largest and most diverse petrified forests on Earth.

Paleontology, the study of fossil life-forms, can be subdivided into specific areas of inquiry: paleozoology, the study of animal fossils; palynology, the study of fossil spores and pollen; and paleobotany, the science associated with the fossil record and evolutionary history of plants. People schooled in these sciences, like scientists of every background, are sometimes asked rather quirky questions; for example, "How many trees, and how many tree species, are there in the United States?"

The response to this specific inquiry, of course, is dependent on a number of variables, including climate, amount of sunshine and rainfall, and type of soil. There is no satisfactory response to the subject of tree count, although one Internet Q and A service is rash enough to state there are nearly 250 billion trees over 1 inch in diameter growing in the fifty states. Less uncertainty is associated with the species count: the US Forest Service is on record stating there are about 865 different types of trees growing in American forests. To date on Earth, however, more than fifty thousand species have been identified.

The wording of the definition of *tree* differs with source, but a widely accepted version is "a tall plant with a thick, wooden stem and many large branches." Trees first emerged during Late Devonian time, 380 million years ago, with *Archaeopteris*, the first recognized "tree" with branches and a trunk. During the Pennsylvanian Period, 70 million years later, extensive growths of ferns, scouring rushes, and scaly-barked trees gave rise to the worldwide deposits of coal that fuel many present-day economies.

Fossil wood is common across the United States, but one locale stands head and shoulders above all others in regard to diversity. With an incredible on-site inventory of more than fifty species, including the first known examples of the native-to-China *Ginkgo biloba*, supposedly the rarest fossil wood, Ginkgo Petrified Forest State Park, in Vantage, is an unprecedented example of a forest of wooden trees transformed into stone.

About 15 million years ago, during the Miocene Epoch, the forests of central Washington grew in a humid and lush environment. Intermittent ash clouds drifting eastward from the tectonically volatile Cascade Range slowly buried magnolia, elm, Douglas fir, cypress, horse chestnut, sassafras, witch hazel, redwood, sweetgum—and ginkgo. Groundwater trickling through these volcanic deposits replaced the organic matter of the dead trees with minerals, principally silica, without destroying identifying characteristics, such as bark and tree rings. Soon, the newly formed varieties of petrified wood were further encased in lava flows, preserving them from the destructive processes of weathering and erosion.

Fast-forward to the end of the Pleistocene Epoch, when as many as eighty cataclysmic floods swept across eastern and central Washington, the result of the periodic collapse of the ice dam that formed Glacial Lake Missoula in northern Idaho. More than 50 cubic miles of soil and bedrock were eroded along the path of the floodwaters, exposing expanses of petrified wood.

The end result of these geologic events is preserved at Ginkgo Petrified Forest State Park. The Trees of Stone Interpretative Trail winds past twenty-two species of fossil logs lying exactly where they were discovered in the 1930s, and a dazzling, world-class collection of highly polished cross sections of petrified wood is on display in the visitor center. This combination of in situ and museum specimens of petrified wood more than adequately supports the claim that Ginkgo Petrified Forest is the most unusual assemblage of stone trees in the world.

Namesake ginkgo leaves growing immediately outside the visitor center are characteristic of present-day foliage at Ginkgo Petrified Forest State Park. Dime for scale.

Inside the visitor center, ginkgo exhibits range from an earth-tone leaf specimen of Ginkgo adiantoides, *a species that thrived until 7 million years ago, to an exquisitely detailed, firewood-sized slab of petrified wood.*

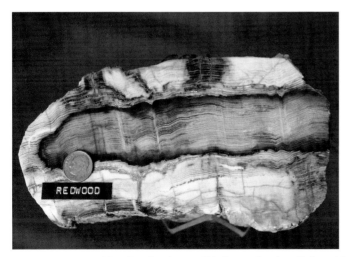

More than four dozen polished examples of petrified wood found within the Ginkgo Petrified Forest State Park are on display in the visitor center, including specimens of redwood (left) and cypress (right). Dimes for scale.

93. Stonerose Interpretive Center and Eocene Fossil Site, Washington

48° 38' 52" North, 118° 44' 21" West

A rare opportunity to smell 50-million-year-old flowers.

Fifty million years ago, during Eocene time, the State of Washington was entering a phase of tectonic activity that would eventually subdivide the land into mountain, desert, and temperate rain-forest habitats. The grandeur of the Cascade Range did not exist, nor did the alpine beauty of the Coast Ranges. The dinosaurs had been extinct for 15 million years, and the evolution of humans was a biological experiment scheduled for the far-ahead future.

In the northeast portion of the state the Okanogan Highlands were being topographically altered by the formation of the Republic graben, a major northwest-trending, fault-bounded depression that became a deep lake covering at least 11 square miles. Over time, deposition by rivers flowing into the lake, combined with ash layers derived from volcanic eruptions that extended across much of the region, filled in the lake with 3,000 feet of sediment that became shale and sandstone. These strata contain a bonanza assemblage of preserved Eocene-age life: stream-transported organic material, in situ lake vegetation, twigs and leaves of shoreline trees and plants, fish, insects, and wind-transported material—all preserved in a contrasting format in a light-toned rock matrix. The richest accumulations are associated with shale strata that split easily into very thin layers; these were deposited rapidly, favoring the burial, rather than the decay, of organic remains. This haven of prehistoric material is open to the public seasonally on a fee basis at the Stonerose Fossil Interpretive Center and Eocene Fossil Site, headquartered at 15 North Kean Street in Republic.

During the Eocene Epoch the North American continent was blanketed by a moist and temperate climate that evolved from the Paleocene-Eocene Thermal Maximum, a warming event that began some 56 million years ago and caused the average global temperature to approach 90 degrees Fahrenheit, compared to a present-day average of 58 degrees Fahrenheit. The temperature gradient from equator to pole was half of that experienced today, sea surface temperatures were as high as 95 degrees Fahrenheit, and the amount of oxygen in the atmosphere had increased almost twofold over a time period of some 20,000 years. Land bridges connected North America with Europe by way of Greenland, Antarctica with Australia, and North America with Asia through the Bering Strait.

A dominating effect of this morphing of land and climate resulted in the spread of temperate forests from pole to pole. Swamp cypress, dawn redwood, beech, and chestnut trees thrived in the Canadian Arctic; palm trees grew in the Dakotas; and fig and magnolia trees flourished in Alaska. The flora of the Pacific Northwest was similar to those that exist today in Central and South America. This was the time when the Stonerose fossil beds were deposited, layers of fossil-laden shale that reveal the secrets of a long-lost land.

At least 450 species of plants are known in the Stonerose fossil beds, quite a few of which have yet to be identified. The specific characteristics are often so finely delineated that paleobotanists can compare them microscopically with modern examples of close affinity. Discoveries include some of the world's oldest well-documented fossil remains of apple, cherry, red raspberry, wild currant, mulberry, and gooseberry plants. Fossils of *Florissantia*, a half-dollar-sized, five-petal flower of an extinct genus of cocoa tree, are so detailed that it has been chosen to be the logo for the fossil site. Leaf specimens have been found with veins, midribs, and stems so exquisitely maintained it would seem they had just fallen to the forest floor.

A *Florissantia* flower in full bloom. The flowers range in size from 1 to 2 inches wide. Fossil fruits and pollen of this Eocene-age plant have also been found. —Courtesy of the Stonerose Interpretative Center and Eocene Fossil Site

The actual fossil dig site is a short walk from the interpretative center. Hammer, chisel, and kneepads are recommended.
—Courtesy of the Stonerose Interpretative Center and Eocene Fossil Site

A March fly from the Stonerose fossil site. "March fly" is perhaps a misnomer, as today this insect is usually found in April and May in the northwestern United States. Quarter for scale. —Courtesy of the Stonerose Interpretative Center and Eocene Fossil Site

This autumn-colored, three-pronged leaf is proof positive that sassafras trees have graced American forests for tens of millions of years. Quarter for scale.
—Courtesy of the Stonerose Interpretative Center and Eocene Fossil Site

94. Glen Lyn Roadcut, West Virginia

37° 21′ 50″ North, 80° 52′ 20″ West

Sea buds and trackways showing changes in terrain and environment.

In the southwestern environs of Mercer County, West Virginia, no more than a hefty stone's throw west of the mountain community of Glen Lyn, Virginia, lies a 1,600-foot-long roadcut along the north side of US 460 that has long provided a sense of wonderment to both fledgling and professional geologists. This rock wall's mixture of steeply dipping Mississippian-age limestone, shale, and sandstone has a story to tell that differs with the teller's perspective.

To the structural geologist, the contorted layers of Bluefield Formation rock mark the transition between two Appalachian Mountain provinces: the folded and faulted strata of the Valley and Ridge Province and the relatively flat-lying rocks of the Appalachian Plateau. To the student of lithology engaged in inch-by-inch examination, the rocks show a definitive change in depositional environment, from one governed by marine conditions that existed for millions of years to one dominated by terrestrial environments, which continue to the present. To the student of fossils, the exposure exemplifies a depository for the uncommonly found blastoid, close cousin to the present-day starfish, sand dollar, and sea urchin.

To the inexperienced eye, blastoid fossils, commonly called "sea buds," look like petrified hickory nuts, but more experienced viewers focus on specifics of their conspicuous geometry. Similar to that of the crinoid, blastoid anatomy is divided into three parts: a primitive root system that anchored the animal to the seafloor; a supporting stem consisting of multitudes of button-like columnals; and a flowerlike crown of feathery food-gathering arms attached to the calyx, the skeletal calcium carbonate cover that enclosed the creature's gut, muscles, nerves, and food grooves (elongated structures used to trap food particles and deliver them to the animal's mouth).

The discovery of a calyx, an interlocking system of fused plates organized in multiples of five, a symmetrical arrangement described as *pentameral*, is the sought-after prize at the Glen Lyn roadcut. Because they are objects of unusual beauty, blastoid calyxes are among the better-known fossils preserved in Mississippian-age rocks throughout the world. Along with as much as 96 percent of their marine brethren, blastoids became extinct in the universal dying-out episode that occurred at the end of the Permian Period. Casts of their columnals, ranging from a fraction of an inch to a full inch in width, are found both in situ in shale fragments and weathered loose in the blankets of scree that cover the base of the rock wall.

A large assortment of blastoid calyxes, some as large as 2.5 inches in diameter, have been removed from the Glen Lyn site, but it's still possible for people possessed with a keen eye and perseverance to find those that have been overlooked. Corkscrew-shaped central structures of the *Archimedes* genus of bryozoan, and fragments of the lacy sheetlike fronds that surrounded the central structure and housed colonies of minute animals, are common in the shale and limestone beds. The valves of firmly rock-embedded brachiopods lie in both concave-down and concave-up orientations, an arrangement suggesting these bilaterally symmetrical seafloor dwellers lived and died in a somewhat tranquil environment. Shell accumulations found entirely in concave-down positions suggest death in a tempestuous environment with storm-generated currents that reoriented each shell into a stable position.

Tetrapod trackways, believed to be the oldest such fossils found in the eastern half of the United States, are present but uncommon in the uppermost strata. The size and shape of forelimb and foot impressions are suggestive of a 2.5-foot-long invertebrate learning to navigate between land and shallow water in the years before the supercontinent Pangaea had finally been assembled.

Two forms of Glen Lyn bryozoan fossils, both belonging to the genus Archimedes: *the highly sought, corkscrew-shaped central structure (right) and less-valued fragments of the lacelike fronds that enveloped it (left). Quarters for scale.*

A pair of hickory nut–sized Glen Lyn Pentremites *blastoid calyxes, showing (right) a down-the-throat view of the centrally located mouth surrounded by five radiating food-gathering grooves and the detail (left) of one food groove (lighter shade of brown). Quarter for scale.* —Courtesy of the Radford University Museum of the Earth Sciences

The Glen Lyn exposure consists of the middle portion of the Bluefield Formation of Middle Mississippian age.

95. Lost River Quarry, West Virginia

39° 03′ 54″ North, 78° 39′ 40″ West

A heckuva place to find a headless variety of every fossil collector's favorite—the trilobite.

Trilobites have been known to science and called by many names for a long time. Seventeenth-century scholars considered them frozen locusts, whereas contemporary enthusiasts have been known to categorize them as butterflies of the sea. Geology students commonly refer to them simply as "bugs." They are found on every continent, define rocks of the Paleozoic Era, and are the most treasured specimens in many a fossil collection. More than twenty thousand species are known, and the count grows every year.

Laterally, trilobite fossils are divided into three distinct segments: the head (cephalon), a central flexible section (thorax), and the tail (pygidium). They can also be divided longitudinally: the thorax and pygidium constitute a central axial lobe sandwiched between two exterior (pleural) lobes. Finding specimens with all three lateral segments (Or all six, depending on how you look at them!) is rare; the more usual find consists of separated body parts.

Semicomplete specimens are present at the Lost River Quarry, a publicly accessible, abandoned roadside excavation reached by exiting US 48 less than 1.5 miles west of Wardensville and traveling 2.3 miles along Old West Virginia 55 to the parking area on the north side of the road. Several different trilobite varieties can be found, but the most common belong to the genus *Phacops*. This well-known Devonian-age trilobite is easily recognized by its large, frog-like eyes and its habit of rolling up into a ball, with the cephalon making a tight seal with the pygidium. This defense mechanism gave it some protection against predators. Many other trilobites had the same habit, but *Phacops* seems to have developed the process to perfection.

The *Phacops* trilobites of the Lost River Quarry belong to the good-and-bad-news category of discovery. The fact that they are generally unrolled is good, but it is not so good that most are headless. Many first-time visitors attribute this decapitated anatomy to predation, a suggestion supported by the rapid evolution and widespread presence of sharks during Devonian time. The idea that a ravenous shark would consume only the head of a several-inch-long trilobite, however, is as unlikely a scenario as that of a starved teenager eating only the olives from a salami, sausage, and olive pizza. Throughout their life trilobites were enclosed in an exoskeleton that sheltered their soft anatomy, much like the shell that protects the soft underbelly of a turtle. Unlike that of the turtle, however, the shell of the trilobite was not a permanent feature; trilobites could only grow by periodically shedding and replacing their exoskeleton, a process called *molting*.

In theory, molting began with the opening of a crack where the cephalon meets the thorax. As the fracture widened the animal gradually wiggled out—ideally with the ease of an egg exiting a chicken—and found a place to hide while a new exoskeleton developed. In practice, however, the process likely progressed haphazardly, creating fractures across both the cephalon and the thorax, forcing the emerging animal to undergo an extended and troublesome trial during which it was subject to opportunistic predation. Most trilobite fossils are molts—discarded exoskeletons—and not the preserved remains of actual animals. The headless specimens at Lost River Quarry are probably molts as well.

Whether rolled up like a pill bug or nakedly exposed while molting, the Lost River Quarry *Phacops* had their ups and downs. They muddled along the path to extinction—one day at a time—until they disappeared completely at the end of the Paleozoic Era.

In addition to trilobites, delicate examples of Trachypora, *an uncommon, branching form of coral that thrived in oxygen-rich waters during the Devonian Period, can be found at the Lost River Quarry.* —Courtesy of www.fossilguy.com

Though not preserved in the best of condition, the subcircular shape and numerous radiating ridges, called costae, *are distinct identifiers for* Rhipidomella, *a common genus of Lost River brachiopods.* —Courtesy of www.fossilguy.com

Six Phacops *exoskeletons. They are classed as molts due to the absence of a cephalon, a common occurrence for this genus of trilobite because of weakness in the animal's shell between its head and thorax.* —Courtesy of www.fossilguy.com

Typically 2 to 3 inches long, Phacops, *like all trilobites, can be divided both laterally and longitudinally into three segments, thus they have a "tri-lobe" anatomy.* —Courtesy of www.fossilguy.com

96. Blackberry Hill, Wisconsin

43° 04′ 15″ North, 89° 24′ 21″ West

Traces of organisms that may have been the first to colonize land.

When referring to the Cambrian chapter of Earth history, it has long been the custom of authors to use words indicative of distinction and unprecedented change, and for good reason. Looking backward from that time period, which spanned 541 to 485 million years ago, a little over 4 billion years of Earth history had transpired, and life had neither taken root in nor left footprints on the embryonic continents. Then almost overnight—at least in the geologic sense—an evolutionary shock wave transformed both the nature of organisms and their fossils. Known as the Cambrian explosion, an unprecedented variety of fauna burst onto the scene, many of which developed exoskeletons for structural support, shelter, and protection. In short, life underwent extensive diversification and began to leave a visible fossil record.

Many aspects of the Cambrian explosion remain controversial: What triggered it? Why did it happen when it did, not later or earlier? Unfortunately, Cambrian rocks are often underlain by an unconformity of gargantuan proportions, forming a gap in the geologic record analogous to a history book containing many missing chapters. Blackberry Hill, 10 miles southeast of Mosinee, is one such "blank" locale where an unconformity, representing 1.5 billion years of lost history, separates Cambrian-age sedimentary rocks from underlying Precambrian igneous rocks. Blackberry Hill is an active fieldstone quarry and closed to the public, but a representative suite of fossil-bearing rock slabs from the site is on daily exhibit at the University of Wisconsin Geology Museum (1215 West Dayton Street) in Madison.

At Blackberry Hill, intersecting traces of trails and pathways and hundreds of dimensional rock impressions are found in Mount Simon Sandstone, deposited in nearshore marine or tidal and river-mouth environments. Two types of trace fossils dominate. The most distinctive and common, named *Climactichnites*, look like 4-inch-wide motorcycle-tire imprints. Paleontologists believe a slug-like organism left the traces as it trolled the shallow water, stirring up the sediment in search of food. The other is reminiscent of tracks left by a long-tailed, meandering, multilegged, horseshoe crab–like animal. Many tracks are tattooed with a myriad of raindrop impressions, rock-solid evidence that they record the first known attempts of animals to migrate out of the seas and walk upon and exploit—but not stay long enough to colonize—the land.

Carnivorous, umbrella-shaped organisms that became stranded ashore during high tides or storms formed the most impressive three-dimensional impressions at Blackberry Hill. Pulsing their gelatinous bodies and repeatedly ingesting and expelling sand in an attempt to regain the sea, these supposed jellyfish were buried and slowly decayed, leaving circular sediment mounds enveloped by trough-like depressions as the only evidence of their existence. The absence of marauding land-based scavengers allowed the imprints to be preserved, and at 3 feet across, they are considered the largest jellyfish-like fossils in the paleontological record.

The Cambrian explosion is a history-altering example of species diversification and evolutionary adaptation. Marine life was moving ever closer to the terrestrial environment, with some forms venturing onshore, at least temporarily, to graze upon bacteria colonies, others to feed upon the grazers. Whatever the needs and whatever the means, life was no longer restricted to the aquatic world.

The organisms represented by the Blackberry Hill fossils in the University of Wisconsin Geology Museum are important. Not only are they unique, but they also offer paleontologists a rare insight into a world populated by marine life experimenting with the ways to establish onshore residences, an evolutionary innovation that would become increasingly more significant throughout the Paleozoic Era.

A 4-inch-diameter imprint of a jellyfish that became stranded on the shore.

Crustaceans scampering across the ripple-marked shoreline of an ancient sea some 520 million years ago left this crisscross display of trackways on this 1.5-to-2-foot-wide slab. Note the elongated grooves associated with the footprints, supposed evidence of a dragging tail. —Courtesy of Steve Uchytil

The 4-inch-wide motorcycle-like track of Climactichnites.

97. Newport State Park, Wisconsin

45° 14′ 17″ North, 86° 59′ 16″ West

An unsurpassed coral coast of alluring architecture and beauty.

The Great Barrier Reef of Australia is both the world's largest coral reef ecosystem and the biggest geographic feature produced by living organisms. Constructed of three thousand separate reef colonies and some one thousand islands, and stretching across 14 degrees of latitudes, the magnitude of its biomass has long given it a reputation as—borrowing a phrase from the financial crisis of 2008—something "too big to fail." Reports that it has lost half of its coral cover in the last three decades, however, indicate this prognostication may not be true. Causes for this reversal are attributed to the effects of global warming, pollution, and an increase in water temperature, among others. As the oceanic environment becomes increasingly stressed, the corals lose their multihued complexion and turn white, a condition known as *bleaching*. While the future of global reef communities remains in doubt, it is important to recall that throughout geologic time marine life has experienced many a roller-coaster ride. One such episode took place during the Silurian Period.

Volcanism was operating in low gear, southern hemisphere glaciers were melting, sea level was rising, and warm, pervasive temperatures blanketed a wide swath of the globe. Across the Lower Peninsula of Michigan the crust of the continent was collapsing into a structure that would ultimately grow to be the Michigan Basin, a 500-mile-wide, bowl-shaped feature that stretches east-west from Detroit to Holland and north-south from the Straits of Mackinac to the Ohio-Indiana border. The shallow waters that inundated this basin, free of the clouding effects of silt and sand, were the perfect incubating environment for the introduction of a new and unusual biologic structure, known today as the "rain forests of the ocean," or reefs.

Reefs are ridgelike forms built through the combined efforts of sedentary, calcareous marine invertebrates performing different tasks. Common inhabitants include solitary and colonial corals living in communal harmony with sponges, clusters of algae, bryozoans, and undulating stalks of crinoids. Brachiopods, trilobites, clams, and snails, all shielded within the confusion of intertwined organisms, complete the roster of life.

Reef communities colonized a broader geographic range during Silurian time than during any other period of history. They became especially prevalent along the northern and western rim of the Michigan Basin. Toward the end of Silurian time, however, conditions changed and the seas became shallower. Exposed to air during low tide and increased salinity, the corals died and reef growth ended—at least for a few million years. Most of these Silurian-age fossil reefs, their pore spaces flush with volumes of crude oil that make Michigan an oil-producing state, are today buried beneath layers of younger sedimentary rock. Some of this intricately fashioned and beautifully preserved coral architecture, however, embedded in glacially scoured carbonate bedrock, is readily visible along the Lynd Point/Fern Trail in Newport State Park on Wisconsin's Door Peninsula.

Two species of tabulate coral—an extinct form of colonial coral distinguished by horizontal, well-developed internal partitions—are present. The *Favosites* genus constructed polygonal cells with perforated walls, resulting in colonial mounds resembling honeycombs or wasp nests, thus the nickname "honeycomb coral." In contrast, the *Halysites* species are commonly labeled "chain corals" because their cell walls are attached side by side in a manner resembling a chain. Fossilized brachiopods, sponges, horn corals, and crinoid parts are found to a lesser degree along the trail. The combination of bygone biology and present-day scenery makes the Newport State Park sector of the Lake Michigan shoreline a true coral coast of unparalleled appeal.

Halysites, *better known as chain coral, has a structure comparable to that of a chain. Each of the small, linked, oval openings housed a coral animal. Penny for scale.*

Looking down the central axis of Zaphrentis, *an extinct, solitary horn coral that lived, side by side with its colonial neighbors, in the Silurian seas 430 million years ago. Penny for scale.*

When seen in cross section, Favosites, *also known as honeycomb coral, has a distinctive hexagonal structure resembling the home of honeybees. Penny for scale.*

An abundance of fossils can be found in the bedding planes of the Cordell Formation dolostone exposed along the Lynd Point/Fern Trail, facing the western shore of Lake Michigan.

98. Two Creeks Buried Forest, Wisconsin

44° 19' 39" North, 87° 32' 41" West

A frigid flora entombed in Pleistocene-age glacial detritus.

Studies of Glaciers, published in 1840, roiled the cloistered halls of academia like a tsunami wave. In this classic treatise Swiss geologist Louis Agassiz presented evidence that a great ice age had once draped Switzerland with a frozen landscape reminiscent of that enveloping Greenland. A year earlier Charles Lyell, the foremost English geologist of the period, introduced the term *Pleistocene* to the scientific community, defining it as an episode of time marking the transition from the recent frigid geologic past to the present. Two giants of intellectual curiosity, one trained in medicine and the other in law, thus set the stage for the proposition that relatively recently global climate had been distinctly cooler than it was during many millions of years of pre-Pleistocene time.

In North America, the Pleistocene has historically been divided into four stages: Nebraskan, Kansan, Illinoian, and Wisconsinan. The Wisconsinan, the youngest of the four and named for glacial deposits extensively studied in the Badger State, stretched from around 25,000 to 10,000 years ago. Many studies during the past century and a half have attempted to subdivide this period of glaciation on an even finer scale. The discovery of the Two Creeks Buried Forest in 1905 helped with this refinement.

Located in the extreme northeast corner of Manitowoc County, the Two Creeks Buried Forest is enveloped by layers of glacial till, but fossilized logs are periodically exposed along the shoreline cliff, depending on storm and coastal erosion conditions. The lowermost glacial deposit, a gray till associated with the Cary substage of glaciation (middle Wisconsinan in age), is overlain by 12-foot-thick, laminated, lake-deposited clays that contain isolated lenses of sand and evidence of tree roots. Next in succession is a thin, organic-rich layer embedded with tree stumps—the remains of an ancient forest. The uppermost layer, a distinctive red till bed and its cache of recumbent logs, is associated with the Valders substage of glaciation (late Wisconsinan in age). Throughout this sequence logs and inclined trunks lie oriented to the southwest, the supposed direction of ice movement in the region.

The 3-inch-thick organic-rich stratum is full of forest litter, consisting of stumps, conifer needles, twigs, peat, pinecones, moss, and terrestrial snails. This mossy forest floor offers the most complete collection of ice age plants known in the region. Identified as a collection of hemlock, pine, and spruce, the logs are well preserved and retain characteristics of their living structure. Analyses of the growth rings indicate the trees thrived in late Pleistocene sunlight an average of sixty years, with one old-timer maintaining its rooted position for eighty-two years. Some of the fossil wood has been radiocarbon dated to 11,850 years. This age provides an absolute date for late glacial activity in the Lake Michigan Basin and supports associative evidence that substage glacial retreats lasted long enough for mature forests to develop.

The mature, boreal-like Two Creeks Buried Forest existed for more than eighty years in the warmer, lake-dominated environment that prevailed after the Cary substage glacier had retreated northward. The advancing Valders glacier bulldozed the forest and, later, melted, flooding and burying it with red till that prevented decomposition. Dubbed by some intellectual wags as a forest that cannot be seen for the lack of trees, the Two Creeks Buried Forest is a world-famous geochronologic site. This state natural area is open to the public, but collecting material is prohibited.

Shielded from view by a field of grass and wildflowers, Two Creeks Buried Forest is better known as a historic fossil site rather than one where Pleistocene-age logs and forest debris can be seen and photographed on a regular basis.

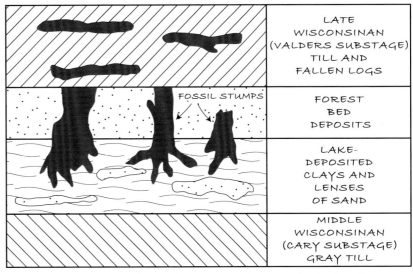

A generalized Two Creeks Buried Forest stratigraphic section, showing the sedimentary layers containing recumbent and in situ logs and stumps and enveloping till beds. (Modified after Prouty 1960.)

The best time to visit the Two Creeks Buried Forest, seen here looking north along the cliff area bordering Lake Michigan, is after the snow has melted but before ground-enveloping vegetation has grown, or after an episode of shoreline erosion caused by an intense lake storm.

99. Fossil Safari, Wyoming

41° 52' 07" North, 110° 40' 34" West

Extinct herring and sardines add pizzazz to a paleontological paradise of fossil fish.

Roughly 48 million years ago the first rays of the morning sun crossed the eastern shore of Fossil Lake, warming the area that would someday become southwestern Wyoming. Dragonflies flitted from lily pad to cattail and mosquitoes swarmed. Luxuriant growths of cypress, fig, and palm trees shaded the shoreline, interspersed with patches of fern. Offshore, predatory crocodiles cruised the depths, the alpha link in a diverse food chain composed of turtles, stingrays, snails, shrimp, and many species of fish.

Fossil Lake was destined to become a fossil graveyard of significance, a locality associated with a variety of detrital conditions. Sluggish streams deposited fine-grained sediment along the shoreline. Clouds of volcanic ash smoothed out topographic irregularities, and a constant "snowfall" of calcium carbonate, precipitated from the chemical-rich water, settled to the lake floor as a slimy ooze. This blanket of physical and chemical sediment was the perfect tomb for incarcerating any organic matter that settled on its surface.

Today the sedimentary rock remnants of this earlier time, a record of a subtropical, freshwater lake ecosystem, are a signature depository of Eocene-age life. These strata, of the Green River Formation, are so rich in fossils that they constitute a *lagerstätte*, a collection of flora and fauna distinguished for its diversity or quality of preservation or, in this case, both. Millions of Green River fish specimens have been added to university and museum collections since they were first catalogued in the 1850s.

Some slabs of Green River Formation contain evidence of a mass-death graveyard. Multiple theories have been forwarded to explain this phenomenon: volcanic eruptions, oxygen depletion, floods, and droughts. A more acceptable explanation envisions an alteration in water chemistry that caused blooms of toxic algae to asphyxiate everything within reach.

In southwest Wyoming, the geologic extravaganza of Fossil Lake fish preservation can be experienced at two locales. Field collecting is not allowed at Fossil Butte National Monument, about 15 miles northwest of Kemmerer on US 30, but there are numerous displays of Green River fossils in the visitor center. Collecting is allowed, however, at Fossil Safari (www.fossilsafari.com), a privately owned quarry about 13 miles northwest of Kemmerer. Here you can keep all the common fish you find—regardless of their size.

Most visitors to the Fossil Safari quarry find enough fossils to satisfy their appetites once they figure out how to use a hammer and chisel to effectively split the highly laminated rock into layers as thin as 0.25 inch. By far the most common find is *Knightia eocaena*, an extinct fish in the herring family. Chosen in 1987 as Wyoming's state fossil, 3-to-6-inch-long specimens of *Knightia* are often found in mass-death clusters. Perhaps their habit of moving in schools—groups swimming together in the same direction in a coordinated manner—led to these mass deaths, as the schools may have unintentionally entered an anoxic zone or a shallow pond that then dried up. *Diplomystus* fossils are also common. Specimens of this 20-to-24-inch-long surface-feeding fish, an extinct relative of the sardine family, are occasionally found with *Knightia* prey lodged in their open jaws, a classic case of predators with eyes bigger than their mouths. In all, more than twelve species of fish have been found at Fossil Safari, along with evidence of leaves, insects, and other forms of life.

Any trip through southwestern Wyoming should include stops at both Fossil Butte National Monument and the Fossil Safari quarry. The Fossil Lake lagerstätte represents a 6-million-year episode when the Eocene Epoch was in full biologic bloom and life was indeed good.

The prominent, firmly set jaw of Diplomystus *suggests this predator hunted smaller fish. Mouth-sized victims are occasionally found preserved in this fish's stomach or stuck in its throat. Dime for scale.*

The degree of detailed fossilization at the Fossil Safari quarry almost defies description. This 8-inch-long specimen of Priscacara liops, *a fish that formed schools like the modern bluegill, features detailed skin, scales, and teeth.*
—Specimen from the collection of the Radford University Museum of the Earth Sciences

Millions of fish specimens at the Fossil Safari quarry remain entombed in hillside exposures (background) and the rubble pile of rock slabs (foreground), both of the Green River Formation.

100. Red Gulch Dinosaur Track Site, Wyoming

44° 27′ 45″ North, 107° 48′ 57″ West

A stone walkway tattooed with the three-pronged prints of flesh-eating dinosaurs.

Most adults have little trouble recalling the plaintive motherly plea directed toward a supposedly irresponsible child to not track footprints into the house. In any abode where a sense of cleanliness and order prevail, footprints on a newly cleaned kitchen floor or a recently vacuumed rug are about as welcome as disease and death. However, when the footprints are found outside and associated with a highly publicized and popular form of extinct life—dinosaurs—their presence is not so disturbing and perhaps even welcome.

Long before European settlers landed in the New World, Indian tribes were aware of dinosaur tracks, although petroglyphs suggest they considered them to be the traces of ancient birds. When Edward Hitchcock released his classic 1858 study of fossil footprints, *The Ichnology of New England*, he perpetuated the concept of their association with birds but did acknowledge that certain trackways might be evidence of creatures with reptile-like tails. Gideon Mantell, the British paleontologist who began the scientific study of dinosaurs, added to the confusion by saying certain dinosaurs left birdlike tracks in their wake by walking on their hind legs. Eventually the concept that a class of unusual vertebrates had evolved and become extinct over a 186-million-year period, the Mesozoic Era, was universally accepted by the scientific community.

For many years, dinosaur studies focused solely on analyzing bones and teeth, skeletal remains that could be assembled into gigantic, awe-inspiring, and attention-gathering museum displays. Tracks and trackways were generally cast aside as curiosities of secondary importance. In the latter decades of the twentieth century, however, this indifference disappeared. Scientists recognized that trackway analyses could provide a rich array of information regarding dinosaur ecology, anatomy, locomotion, and behavior. Not at all uncommon in the United States, publicly accessible dinosaur track sites are often protected behind barriers or within "do not touch" displays under roofs. The Red Gulch Dinosaur Track Site, however, 5 miles south of US 14 and between the communities of Shell and Greybull, is an exception.

Numbering in excess of 1,100, the Red Gulch imprints are preserved in a cream-colored, ripple-marked limestone of the 167-million-year-old Sundance Formation. Nearly 125 discrete trackways, ranging in length from two to forty-five steps, have been catalogued. The majority of the tracks are three-digit shapes, longer than they are wide, accentuated by sharp claw marks. These characteristics are generally associated with the theropod class of short-fore-limbed carnivorous dinosaurs that walked or ran on strong hind legs.

Statistical analyses of hundreds of prints suggest only one type of dinosaur populated the Red Gulch area—individuals less than 7 feet tall and weighing no more than several hundred pounds. By comparing wind-generated ripple marks with the subparallel trackway paths, scientists conclude the dinosaurs walked at a pace of about 4 miles per hour, to the south or southeast, into the wind and perhaps in pack formation—most certainly on the quest for food.

When these dinosaurs frequented the Red Gulch area, it was a tidal flat along the shoreline of the Sundance Sea, the marine invasion that covered vast sections of the western United States during Middle Jurassic time. Although the species have yet to be identified, during their short period of time on Earth they left a distinctive digital record that is arguably second to none.

View across and along the "ballroom," the central portion of the Red Gulch trackway, containing hundreds of well-preserved tracks. The overlying darker-brown formation contains specimens of Gryphaea nebrascensis, a genus of extinct oyster that lived in a shallow marine environment, possibly as large colonies.

Fossil documentation of heavy traffic during a Mesozoic Era rush hour.

The deeply embedded nature of this three-toed footprint suggests it was made by a fully grown, mature adult. Half-dollar for scale.

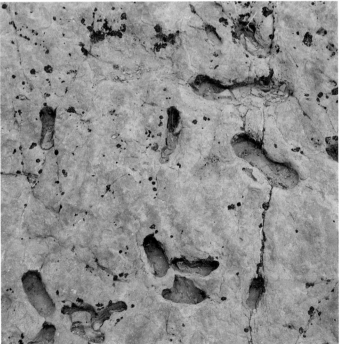

101. Dinosaur Center, Wyoming

43° 38′ 35″ North, 108° 11′ 59″ West

***Archaeopteryx* and a basketball court–sized display of dino skeletons gives definition to the Age of Reptiles.**

Every spring thousands of trekkers take to field and forest to engage in an activity that requires nothing more than a good pair of binoculars, a well-illustrated field guide, and a sense of unrelenting enthusiasm that lasts until fatigue and sunset direct them home. Bird-watching, or birding, is a recreational hobby that some fifty million Americans undertake, intent on observing a maximum number of the more than eight hundred bird species that now and then call North America home. Captivated by the swooping aerial display of the hummingbird, the mimicking call of the mockingbird, or the burst of color that adorns the painted bunting, conventional birders see the objects of their interest as benign, chirping creatures adapted to a sociable, neighborly, and beneficial life in harmony with nature. Birders who trace the evolutionary record of birds back in time, however, realize that early avian ancestors descended from a life of dominance and competition. This less-benign heritage begins with *Archaeopteryx*, an exotic, crow-sized, feathered vertebrate—the granddaddy of all birds.

After examining the blackish blur of a fossil discovered embedded in a slab of Late Jurassic limestone southwest of the community of Solnhofen, Germany, by an unknown quarryman in 1860, Hermann von Meyer, a German paleontologist, described it as a wing feather 2.3 inches long, asymmetrical in shape, and partridge-like in appearance. Baptized with the name *Archaeopteryx*—from the Greek words *archaios*, meaning "ancient," and *pteryx*, meaning "wing"—it became the holotype for the earliest known bird, even though an actual skeleton had yet to be found. Within a year, however, and before the prototype feather could be formally described in the scientific literature, a birdlike skeleton was discovered in the same 150-million-year-old limestone. Although the head and neck were missing, contemporary textbooks commonly describe this fossil as having belonged to the oldest known bird. In his sixth edition of *On the Origin of Species*, Charles Darwin mentioned it as "that strange bird, *Archaeopteryx*, with a long lizard-like tail, bearing a pair of feathers on each joint and with its wings furnished with two free claws" (Feduccia 2012). Today the twelve *Archaeopteryx* specimens that exist are scattered in museums around the world. The Thermopolis specimen, the crown jewel of the Wyoming Dinosaur Center (110 Carter Ranch Road) in Thermopolis, is the only one on display in North America.

Although *Archaeopteryx* is associated with the phrase "first bird," paleontologists agree it is actually a hybrid fossil with both avian and reptilian features. Like modern-day birds, *Archaeopteryx* possessed wings with flight feathers, a birdlike beak, a partially reversed first toe, and a wishbone. In contrast, a handful of teeth, a long bony tail, and three claws jutting out from the midpoint of each wing are evidence that it was related to theropods, the class of carnivorous dinosaurs that includes the 40-foot-long *Giganotosaurus*, the 3-ton *Albertosaurus*, and the infamous predator *Tyrannosaurus rex*. The anatomical details of *Archaeopteryx* support the concept that birds with reptilian traits evolved from dinosaurs with avian characteristics—body-structure change that strongly supports Darwin's theory of evolution.

Additional attractions in the Wyoming Dinosaur Center complex include the Walk Through Time, an introduction to the evolution of ancient life from primitive forms to the rise of reptiles, and the Hall of Dinosaurs, where displays include the *Supersaurus* "Jimbo," one of the largest dinosaurs ever mounted; "Stan," a 35-foot-tall *T. rex*; *Triceratops*, the Wyoming state dinosaur; a nest of eighteen dinosaur eggs; and thirty more exhibits of fossil skeletons.

A nest of 6-inch-long, 77-million-year-old eggs modeled after the famous Egg Mountain, Montana, discovery of 1893, the first site where dinosaur eggs were found. The mother, a reptile in the genus Maiasaura, *incubated her developing offspring by covering them with layers of rotting vegetation.*

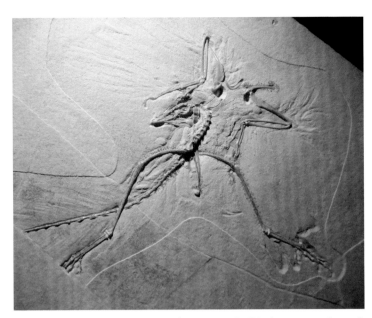

The Wyoming Dinosaur Center's Archaeopteryx *fossil is the most complete and well-preserved example found to date. Possessing both reptilian and avian features, it is recognized as the ancestor of all birds.*

The faintly visible front and rear limb feathers of Microraptor, *a crow-sized, four-winged dinosaur of the Early Cretaceous Period, support the idea that birds and dinosaurs have an evolutionary relationship. Analyses of fossilized pigment cells suggest* Microraptor *wings were thick, glossy, and iridescent.*

Glossary

Acadian orogeny. The second of the three Paleozoic-age phases of mountain building that were instrumental in the development of the Appalachian Mountains.

adaptive radiation. The process by which organisms diversify rapidly into a multitude of new forms under conditions of environmental change.

agnathids. A class of vertebrate organisms, a modern example being the lamprey.

algae. Primitive, chiefly aquatic plants that lack stems, roots, and leaves but usually contain chlorophyll.

Alleghanian orogeny. The third of the three Paleozoic-age phases of mountain building that were instrumental in the development of the Appalachian Mountains.

ammonite. The coiled and chambered shell of a class of extinct mollusk.

ammonoid. Any extinct cephalopod characterized by an external shell that is symmetrical and coiled in a plane.

amniotic. Relating to the amnion, a thin membrane forming a closed sac around the embryos of reptiles, birds, and mammals and containing the amniotic fluid.

anoxic. An environment lacking in oxygen.

anticline. A convex-upward fold in layered rock commonly caused by compressive tectonic forces.

Appalachian Plateau. The physiographic province that forms the northwestern part of the Appalachian Mountains, stretching from New York to Alabama and largely composed of coal fields of Pennsylvanian-age rock.

ash. Extremely fine-grained material derived from a volcanic eruption.

asteroid. Celestial bodies having diameters between one and several hundred miles and orbits that lie chiefly between Mars and Jupiter.

badland. Topography with little or no vegetation and characterized by a rugged, high-relief terrain formed within unconsolidated or poorly cemented clay or silt.

basalt. A dark-colored, fine-grained, extrusive igneous rock often seen in the form of lava flows but also in the form of small intrusive bodies, such as dikes or sills.

basement. The undifferentiated complex of rocks, often a combination of igneous and metamorphic types, that underlies the rocks of interest in an area.

Basin and Range Province. A region that extends through much of the southwestern United States and is characterized by numerous north-south-oriented valleys bounded by mountain ranges.

bed. A layer of sedimentary rock bounded above and below by well-defined bedding planes.

bedding plane. The surface of separation between any two distinct beds of sedimentary rock.

bedrock. A general term for the solid rock that underlies soil or other superficial material.

belemnite. The cigar-shaped internal shell of various extinct cephalopods related to the cuttlefish.

benthic. Relating to or occurring at the bottom of a body of water.

biomass. The total mass of organisms in a given area or volume.

biota. The total amount of animal and plant life found in any particular area.

bivalve. Having a shell composed of two distinct and usually movable valves that open and shut.

blastoid. An extinct type of echinoderm, often called "sea buds," with fossils that look like small hickory nuts.

body fossils. The remains of actual organisms, such as shells, bones, and teeth.

boulder. A rock mass having a diameter greater than 10 inches.

brachiopod. Any solitary marine invertebrate characterized by two symmetrical valves that are commonly attached to a substratum.

bryozoan. A small invertebrate characterized by colonial growth and a calcareous skeleton.

butte. An isolated, flat-topped, steep-sloped hill with a summit smaller in extent than that of a mesa.

calcareous. Consisting of calcium carbonate.

calcite. A common mineral composed of calcium carbonate.

calcium carbonate. A whitish crystalline compound found in limestone, chalk, and many shells.

calyx. The cup-shaped skeletal cover of an echinoderm.

carbonate. A sedimentary rock composed of appreciable amounts of limestone, chalk, or dolomite.

carbonization. The process of reducing or converting fossil remains to carbon.

carnivorous. Belonging to or pertaining to flesh-eating or predatory animals.

cast. Secondary rock or mineral material that fills a natural mold, forming a reproduction of the external details of a fossil shell, skeleton, or other organic structure.

catastrophism. The theory that changes in Earth's crust during geologic history have resulted mainly from violent and unusual events.

cephalon. The segmented head region of an exoskeleton of a trilobite bearing the eyes and mouth.

cephalopod. A marine mollusk characterized by a head with a mouth surrounded by part of the foot modified with tentacles.

chalk. A soft, fine-grained rock composed almost entirely of the calcareous remains of microorganisms or algae.

chert. A hard, fine-grained rock composed entirely of silica.

chitinous. Consisting of chitin, a resistant organic compound common to various invertebrate skeletons.

cinder cone. A conical volcanic feature composed of fragments of igneous rock emitted during an explosive eruption.

clay. A particle of sediment less than 0.00016 inch (0.004 millimeter) in diameter.

coal. A combustible rock formed by the compaction of altered plant remains.

columnal. One of the numerous individual plates that make up the column, or stem, of an echinoderm.

concretion. A compact mass of mineral matter formed by precipitation from an aqueous solution about a nucleus, such as a leaf, bone, or fossil.

conglomerate. A coarse-grained sedimentary rock composed of rounded pebbles, cobbles, and boulders held together by a fine-grained matrix.

coprolite. Fossilized excrement.

corals. Marine organisms, generally colonial in nature, characterized by calcareous skeletons massed in a variety of shapes and often forming reefs or islands.

core. The central zone of Earth's interior below the depth of 1,800 miles. It is divided into an inner, solid subcore and an outer, fluid subcore.

crater. A bowl-shaped, rimmed feature forming the summit of a volcano.

crinoid. Marine organisms characterized by feathery radiating arms and a stalk by which they are attached to a surface.

cross bed. A single bed inclined at an angle to the main planes of stratification.

crust. The outermost shell of Earth, which varies in thickness from 6 to 25 miles.

crustacean. A class of aquatic organisms that includes lobsters, crabs, and shrimp.

cyanobacteria. Small, generally unicellular aquatic organisms that can manufacture their own food. They are believed to be the oldest known fossils, at more than 3.5 billion years old.

cycad. Any seed-bearing plant resembling a palm tree but topped by fernlike leaves.

differential weathering. Weathering that occurs at different rates as a result in variations in the composition and resistance of a rock.

dissolution. The dissolving of rock through chemical alteration.

dolomite. A variety of limestone rich in magnesium carbonate.

dugong. An herbivorous marine mammal.

earthquake. Ground displacement caused by the sudden release of built-up stress in the crust of the Earth.

echinoderm. A radially symmetrical class of marine invertebrates that includes starfish and sea urchins.

epicontinental. Pertaining to something situated in the interior of continent, such as a sea.

epoch. The formal geochronologic unit longer than an age and shorter than a period.

era. The formal geochronologic unit next in order of magnitude below an eon.

erosion. The removal of soil and weathered rock by wind, water, ice, or gravity.

erratic. A rock fragment carried by glacial ice and deposited on bedrock of a different lithology than the outcrop from which it was derived.

exoskeleton. A supportive and protective framework that lies outside the body tissue of an animal.

fault. A break in a rock along which there has been movement of one side relative to the adjacent side.

fauna. The collective group of animals found in any particular region.

fissure vent. A crack or fissure that forms where a volcanic conduit meets the surface of the Earth.

flint. A hard, fine-grained form of quartz that sparks when struck with steel.

flora. The collective group of plants found in any particular region.

flowstone. A deposit of calcium carbonate formed by water flowing on the walls or floor of a cave.

fold. A bend in rock strata or bedding planes.

foraminifera. Very small marine organisms that form calcareous shells, which are useful for age dating sedimentary rock.

formation. The basic cartographic rock unit in geologic mapping. A formation has easily recognizable boundaries that can be traced in the field.

fossil. The remains or evidence of preexisting life.

gastropod. A class of mollusk with a body contained in an asymmetrical, helically coiled shell.

genera. A category of plant and animal classification intermediate in rank between family and species.

geochronology. The science of dating and determining the time sequence of events of Earth history.

geomorphology. The study of the classification, description, origin, and development of landforms.

ginkgo. A type of gymnosperm tree with fan-shaped or regularly bifurcated leaves.

glacial polish. A bedrock surface smoothed by glacial abrasion.

glaciation. The formation of glaciers and the effects they have on a landscape.

glacier. A mass of ice that moves outward in all directions due to the stress of its own weight.

Gondwanaland. The continent formed during the early Paleozoic Era in the southern hemisphere by the amalgamation of several preexisting microcontinents.

graben. An elongate trough bounded on both sides by faults that dip toward the interior of the trough.

granite. A coarse-grained, light-colored, acidic intrusive rock rich in quartz and potassium feldspar.

graptolites. Small, colonial, aquatic, extinct animals that have been preserved mainly as carbonaceous impressions on the surfaces of shale rock.

gymnosperm. A plant whose seeds are commonly enclosed in cones, such as pine, fir, and spruce.

herbivore. A plant-eating animal.

hogback. A ridge with a distinct, sharp summit and steep slopes of equal inclination.

holotype. A single example of a plant or animal used to formally describe a new class of organism.

hominid. A primate of the family of modern man.

horn coral. A solitary coral whose miniature, cone-shaped skeletal remains are characterized by a wrinkled surface.

humanoid. Having an appearance or character resembling that of a human being.

hydrocarbon. Any gaseous, liquid, or solid compound consisting of carbon and hydrogen, crude oil being a common example.

Iapetus Ocean. Considered the predecessor to the present-day Atlantic Ocean, this ocean was destroyed by the amalgamation of northern hemisphere microcontinents during the formation of the landmass of Laurasia.

ice field. An extensive mass of ice distributed across a mountain region and covering all but the highest peaks and ridges.

ichnofossil. A sedimentary structure consisting of a fossilized track, trail, tube, boring, or tunnel resulting from the life activities of an animal.

igneous. A class of rock formed by the crystallization of magma.

impression. The form, shape, or indentation made on a soft sedimentary surface by an organic or inorganic structure, such as a fossil shell or the surface of a leaf.

index fossil. A fossil that identifies and dates the strata or succession of strata in which it is found.

in situ. Meaning "in its original place."

interbedded. Said of beds lying between or alternating with others of different character.

intrusion. A mass of igneous rock that intruded preexisting rock.

invertebrate. A class of animal that does not have a backbone.

isotope. One of two or more species of the same chemical element.

karst. Topography formed on and within a carbonate rock foundation and characterized by caves, sinkholes, and other dissolution features.

Kaskaskia Sea. An epicontinental sea that invaded North America during middle Paleozoic time.

lagerstätte. A rock deposit that has an abundance and variety of fossils, often preserved in fine detail and showing soft body parts.

lamination. The thinnest recognizable layer of original deposition in a sediment or a sedimentary rock.

lampshell. A synonym of brachiopod.

land bridge. A terrestrial connection between continents that permits the migration of organisms.

Laurasia. An ancient landmass formed in the northern hemisphere by the amalgamation of several preexisting microcontinents. The supercontinent Pangaea was formed by the joining of Laurasia and Gondwanaland.

Laurentia. A former microcontinent that today forms the geologic heart of North America. Also known as proto–North America.

lava. Molten rock that erupted onto Earth's surface.

lava dome. A dome-shaped mountain formed by the extrusion of many layers of very fluid lava.

lens. A convex geologic deposit bounded by converging surfaces that are thick in the middle and thin toward the edges.

limestone. A sedimentary rock composed chiefly of calcium carbonate. Typically forms in lakes and warm, shallow seas.

limonite. A widely occurring, yellowish-brown to black natural iron oxide used as an ore of iron.

magma. Molten or partially molten rock found in the interior of Earth. Called lava when it erupts onto the surface.

magnetite. The mineral form of black iron oxide.

mammoth. An extinct elephant once found throughout the northern hemisphere.

mantle. The zone of the Earth that lies below the crust and above the core.

marble. The metamorphic form of a carbonate rock.

marl. Loose, earthy material consisting chiefly of clay and calcium carbonate, formed under marine or freshwater conditions.

massive. Said of rock with homogenous texture and lacking layering, fractures, or other discontinuities.

mastodon. One of a group of extinct, elephant-like mammals widely distributed in the northern hemisphere during Oligocene through Pleistocene time.

matrix. The finer-grained material enclosing or filling the interstices between the larger grains or particles of sediments or sedimentary rock.

megafossil. A fossil large enough to be studied without the aid of a microscope.

megascopic. Visible to the naked eye.

mesa. An isolated, flat-topped, steep-sloped hill with a summit more extensive than that of a butte.

metamorphic. Pertaining to the process of metamorphism or to its result.

metamorphism. Mineral, chemical, and physical changes to solid rock due to being subjected to conditions characterized by environments of high temperatures and high pressures.

meteorite. A celestial mass of rock that fell to Earth.

microfossil. A fossil too small to be studied without the aid of a microscope.

microtektite. A small (less than 1 millimeter in diameter), rounded, dark-colored body of silicate glass believed to be the product of a large, hypervelocity meteorite impact on sediments, although it may have a lunar origin.

mold. An impression made in the surrounding earth or rock material by the exterior or interior of a fossil shell or other organic structure.

mollusk. A solitary invertebrate characterized by a non-segmented body that is bilaterally symmetrical and by a radially or biradially symmetrical mantle and shell.

monolith. A large, upstanding, and generally unfractured mass of rock.

moraine. A ridge of till that accumulated along the front and sides of a glacier.

mudflow. A very fluid, flowing mass of fine-grained sediment.

mudstone. A massive, bedded, indurated mud having the texture and composition of shale but lacking the fine layering that is characteristic of shale.

normal fault. A fault along which rock on one side has moved downward relative to rock on the other side.

obsidian. A dark volcanic glass that fractures in a conchoidal pattern.

oil seep. The emergence of migrating liquid petroleum at the surface of the Earth.

opal. A hydrated and amorphous form of silica.

opalized. Said of a fossil that was converted into a form of opal or chalcedony.

orogeny. The general tectonic process in which mountains are formed.

ostracod. Any aquatic crustacean characterized by a bivalve, generally a calcified carapace with a hinge along the dorsal margin.

outcrop. The part of a formation or geologic structure that is exposed at the surface of the Earth.

paleoclimate. The climate of a given interval of time in the geologic past.

paleogeography. The geography of a given interval of time in the geologic past.

paleontology. The study of life of the geologic past, as based on fossil plants and animals.

Pangaea. The supercontinent that formed during the late Paleozoic Era through the merger of Gondwanaland and Laurasia.

Panthalassa Sea. The ancient ocean that surrounded the supercontinent of Pangaea.

peat. An unconsolidated deposit of semicarbonized plant remains in a water-saturated environment, such as a bog.

pebble. A generally rounded stone between 0.16 and 2.5 inches (4 and 64 millimeters) in diameter.

pedicle. A variably developed, fleshy or muscular appendage of a brachiopod that attaches the animal to substratum.

pelecypod. A class of bivalve mollusk.

period. The geochronologic unit lower in rank than an era and higher than an epoch.

permineralization. A process of fossilization whereby the original hard parts of an animal have additional mineral matter deposited in their pore spaces.

petroglyph. A rock image created by removing part of the rock surface by incising, picking, carving, or abrading.

photosynthesis. A biologic process whereby carbon dioxide is converted to organic matter in the presence of light.

phylum. The primary taxonomic division of the animal kingdom.

placoderm. A class of jawed vertebrates characterized by heavy dermal armor.

plankton. Aquatic organisms that drift, or swim weakly.

plate tectonics. The theory that posits that most large-scale features of Earth form through the relative movement and interaction of the rigid plates composing the outer portion of the planet.

precipitation. Water that falls to the surface from the atmosphere as rain, snow, hail, or sleet.

principle of faunal succession. The concept stating that fossil assemblages in sedimentary strata succeed one another in a regular and determinable order.

principle of fossil correlation. The concept stating that sequences of strata containing similar collections of fossils are of similar age.

principle of superposition. The concept stating that in a vertical sequence of undisturbed layered rock, the oldest layer is at the base and the youngest on top.

proboscidean. The order of mammals that is characterized by a trunk, or proboscis, and includes the elephant.

protocontinent. A landmass that in the course of geologic time could develop into a continent.

pygidium. The fused and segmented tail piece of the exoskeleton of a trilobite.

pyrite. A common pale-bronze or brass-yellow cubic mineral composed of iron and sulfur.

quartz. A common rock-forming mineral composed of silica and oxygen; typically clear but can come in many colors due to the inclusion of certain elements.

quartzite. A quartz-rich sandstone that has been metamorphosed.

radiocarbon dating. A method for determining the age of an object containing organic matter by using the properties of a radioactive isotope of carbon.

reef. A mound-like structure built principally of and by calcareous marine organisms.

Rheic Ocean. The ocean that separated Gondwanaland and Laurasia, two major paleocontinents.

rift. A long and narrow trough bounded by normal faults that completely penetrate the lithosphere. A rift forms as a continent is pulled apart by tectonic extension.

riparian. Pertaining to the bank of a river, pond, or small lake.

ripple mark. An undulatory rock surface composed of small-scale ridges and hollows, formed as wind or flowing water moved sediment around before it hardened into rock.

Rodinia. The supercontinent that formed toward the end of the Proterozoic Eon; the predecessor of the supercontinent Pangaea.

rugose. Coarsely wrinkled, uneven, and rough, as on the surface of horn coral fossils.

sand. A detrital rock fragment or mineral particle having a diameter in the range of 0.0025 to 0.08 inch (0.0625 to 2 millimeters).

sandstone. A sedimentary rock composed of rounded, sand-sized sediment held together by a naturally occurring cement.

scree. A heap of broken rock fragments and the steep slope consisting of such fragments.

sediment. Solid fragmental material that originates from the weathering of rocks and is transported or deposited by air, water, or ice.

sedimentary. A class of rock formed by the deposition and cementation of sediment.

sedimentation. The deposition of sediment by wind, water, or ice.

shale. A sedimentary rock composed of clay- and silt-sized sediment and having a tendency to break along thin, parallel planes.

siderite. A yellowish-brown mineral that is a valuable ore of iron.

silica. Silicon dioxide, a chemical compound composed of silicon and oxygen. The chief constituent of quartz.

silt. A sedimentary particle ranging in size between 0.00016 and 0.0025 inch (0.004 and 0.0625 millimeter) in diameter.

siltstone. A sedimentary rock composed of silt-sized sediment with a texture that is intermediate between sandstone and shale.

sinkhole. A circular topographic depression in an area of karst.

slate. Shale that has been metamorphosed.

soda straw. A tabular stalactite that resembles a drinking straw and has the diameter of a drop of water.

speleothem. Any of various types of mineral deposits that form in caves.

spicules. Minute, needlelike, calcareous or siliceous bodies that support the tissues of various invertebrates, such as a sponge.

stalactite. A speleothem that hangs from the ceiling of a cave and develops as minerals precipitate from water that drips onto the floor.

stalagmite. A speleothem that grows upward from the floor of a cave and develops as minerals precipitate from water that drips onto the floor.

strata. Tabular layers of sedimentary rock separated from each other by a distinctive bedding plane.

stratigraphy. The branch of geology that studies sedimentary and layered volcanic rocks and interprets them in terms of a general timescale.

stromatolite. An organic sedimentary structure produced by sediment being trapped or bound as a result of the growth of microorganisms, principally cyanobacteria.

stromatoporoid. A general name for any group of extinct, sessile benthic marine organisms of uncertain biologic affinities.

suite. Any succession of related things, such as a fossil suite.

supercontinent. A landmass, such as Pangaea or Rodinia, formed periodically by the amalgamation of all the existing continents.

Taconica. An ancient arc of islands that was added to the core of North America during the early Paleozoic Era.

Taconic orogeny. The first of the three Paleozoic-age phases of mountain building that were instrumental in the development of the Appalachian Mountains.

talus. An assemblage of angular rock fragments lying at the base of a cliff or steep slope.

tectonics. Pertaining to the forces involved in the development of the broad architecture of Earth's crust, such as ocean basins, mountains, folds, and faults.

terrain. A region of the Earth that is considered a physical feature, such as the Great Plains.

terrane. A body of rock bounded by faults and characterized by a geologic history that differs from adjacent terranes.

tetrapod. An animal with four limbs.

theca. An echinoderm skeleton consisting of calcareous plates that enclose the body and internal organs.

thorax. The middle part of the exoskeleton of a trilobite.

till. Sediment deposited directly by and underneath a glacier without subsequent reworking by meltwater and consisting of a mixture of clay, silt, sand, and boulders of varying size and shape.

Tippecanoe Sea. An extension, or arm, of the Iapetus Ocean, which covered much of what is today North America during the Ordovician and Silurian Periods.

trace fossil. An alternative term for ichnofossil.

trackway. A continuous series of tracks formed by a single organism.

trilobite. An extinct marine arthropod characterized by a three-lobed exoskeleton divisible both longitudinally and latitudinally.

tsunami. A long-wavelength sea wave produced by a large-scale disturbance of the ocean floor, such as a submarine earthquake.

tundra. A treeless, level, or gently undulating plain characteristic of arctic and subarctic regions.

turbidity. The state and condition of opaqueness or reduced clarity of a fluid due to the presence of suspended matter.

turbidity current. A density current in air, water, or another fluid caused by different amounts of matter in suspension.

unconformity. A gap in the geologic record, generally within a series of sedimentary rocks, in which a sequence of strata is missing either because of erosion or because no sediment was being deposited at that time.

uniformitarianism. The principle that the geologic forces of the past differ neither in kind nor in energy from those now in operation.

Valley and Ridge Province. The region of the Appalachian Mountains that consists of elongate, parallel ridges and valleys underlain by thick sequences of folded sedimentary rock of Paleozoic age.

valve. One of the distinct and usually articulated pieces that make up the shell of certain invertebrates.

vertebrate. Any animal with a backbone or spinal column.

volcanic. Pertaining to the activities, structures, or rock types associated with a volcano.

weathering. The physical and chemical breakdown of rocks at Earth's surface due to exposure to the atmosphere and hydrosphere.

Western Interior Seaway. The sea that extended south from the Arctic Ocean to the Gulf of Mexico and 600 miles from the embryonic Rockies east to the Appalachians, engulfing the central portion of North America during the Cretaceous Period.

woolly mammoth. A species of mammoth that lived during the Pleistocene Epoch.

zircon. A mineral composed of zirconium silicate and commonly employed in the absolute age-dating of rocks because of its resistance to weathering and erosion.

zooid. An individual animal in a colony of animals.

zooplankton. Animals that live all or part of their life suspended and drifting in fresh or salt water.

References

Introduction

Hooke, R., J. Martyn, and J. Allestry. 2015. *Micrographia*. Reprint. Delhi, India: Facsimile Publisher.

Jackson, J. A., ed. 1997. *Glossary of Geology*. 4th ed. Alexandria, VA: American Geological Institute.

Phillips, J. 2011. *Memoirs of William Smith, LL. D. Author of the "Map of the Strata of England and Wales": By His Nephew and Pupil*. Reprint. New York: Cambridge University Press.

Rudwick, M. J. S., ed. and trans. 1997. *Georges Cuvier, Fossil Bones, and Geological Catastrophes: New Interpretations of the Primary Texts*. Chicago: University of Chicago Press.

1. Double Roadcut, Alabama

Copeland, C. W. 2007. *Curious Creatures in Alabama Rocks: A Guide Book for Amateur Fossil Collectors*. Geological Survey of Alabama Circular 19.

Hess, H., W. I. Ausich, C. E. Brett, and M. J. Simms 1999. *Fossil Crinoids*. New York: Cambridge University Press.

Lacefield, J. 2013. *Lost Worlds in Alabama Rocks: A Guide to the State's Ancient Life and Landscapes*. Tuscaloosa, AL: Alabama Museum of Natural History.

McKinney, F. K. 1991. *Bryozoan Evolution*. Chicago: University of Chicago Press.

2. Minkin Footprints, Alabama

Buta, R. J., and D. C. Kopaska-Merkel. 2016. *Footprints in Stone: Fossil Traces of Coal-Age Tetrapods*. Tuscaloosa, AL: University of Alabama Press.

Buta, R. J., J. C. Pashin, N. J. Minter, and D. C. Kopaska-Merkel. 2013. "Ichnology and Stratigraphy of the Crescent Valley Mine: Evidence for a Carboniferous Mega-Tracksite in Walker County, Alabama." In *The Carboniferous-Permian Transition: New Mexico Museum of Natural History and Science Bulletin* 60, edited by S. G. Lucas, W. A. DiMichele, J. E. Barrick, J. W. Schneider, and J. A. Spielmann, 42–56.

Kopaska-Merkel, D. C., and R. J. Buta. 2012. *Field-Trip Guidebook to the Steven C. Minkin Paleozoic Footprint Site*. Birmingham, AL: Alabama Paleontological Society.

MacDonald, J. 1995. *Earth's First Steps: Tracking Life Before the Dinosaurs*. Boulder, CO: Johnson Books.

Minkin, S. C. 2000. "Pennsylvanian Vertebrate Trackways." *Alabama Geological Society Newsletter* 13 (4): 7–10.

3. Moscow Landing, Alabama

Copeland, C. W. 1963. *Curious Creatures in Alabama Rocks: a Guide Book for Amateur Fossil Collectors*. Geological Survey of Alabama Circular 19.

Jones, R. W. 2014. *Foraminifera and Their Applications*. Cambridge, UK: Cambridge University Press.

Lacefield, J. 2013. *Lost Worlds in Alabama Rocks: A Guide to the State's Ancient Life and Landscapes*. Tuscaloosa, AL: Alabama Museum of Natural History.

Smith, C. C. 1997. "The Cretaceous-Tertiary Boundary at Moscow Landing, West-Central Alabama." In *Gulf Coast Association of Geological Societies: Transactions*, vol. 47, edited by W. W. Craig and B. Kohl, 533–539.

4. Coyote Lake, Alaska

Conner, C. 2014. *Roadside Geology of Alaska*. Missoula, MT: Mountain Press Publishing.

Harriman, E. H., and C. H. Merriam. 2010. *Alaska: Geology and Paleontology*. Charleston, SC: Nabu Press.

Sunderlin, D., G. Loope, N. E. Parker, and C. J. Williams. 2011. "Paleoclimate and Paleoecological Implications of a Paleocene-Eocene Fossil Leaf Assemblage, Chickaloon Formation, Alaska." *Palaios* 26 (6): 335–345.

Triplehorn, D. M., D. L. Turner, and C. W. Naeser. 1984. "Radiometric Age of the Chickaloon Formation of South-Central Alaska—Location of the Paleocene-Eocene Boundary." *Geological Society of America Bulletin* 95 (6): 740–742.

5. Indian Gardens, Arizona

Chronic, H. 1983. *Roadside Geology of Arizona*. Missoula, MT: Mountain Press Publishing.

Dunbar, C. O., and G. E. Condra. 1932. *Brachiopods of the Pennsylvanian System in Nebraska*. Nebraska Geological Survey Bulletin 5.

Grinnell, R. S., and G. W. Andrews. 1964. "Morphologic Studies of the Brachiopods Genus *Composita*." *Journal of Paleontology* 38 (2): 227–248.

Lutz-Garihan, A. B. 1976. Composita subtilita *(Brachiopod) in the Werford Megacyclothem (Lower Permian) in Nebraska, Kansas and Oklahoma*. University of Kansas Paleontological Contributions Paper 81.

6. Petrified Forest, Arizona

Ash, S., and D. D. May. 1990. *Petrified Forest: The Story Behind the Scenery*. Petrified Forest, AZ: Petrified Forest Museum Association.

Daniels, F. J., and R. D. Dayvault. 2006. *Ancient Forests: A Closer Look at Fossil Wood*. Grand Junction, CO: Western Colorado Publishing.

Hickey, L. J. 2010. *The Forest Primeval: The Geologic History of Wood and Petrified Forests*. New Haven, CT: Yale Peabody Museum.

Lubick, G. M. 1996. *Petrified Forest National Park: A Wilderness Bound in Time*. Tucson, AZ: University of Arizona Press.

7. Devils Backbone, Arkansas

Eckert, A. W. 2000. *Earth Treasures: The Southwestern Quadrant*. Lincoln, NE: iUniverse.com, Inc.

McFarland, J. D., et al. 1979. *A Guidebook to the Ordovician-Mississippian Rocks of North-Central Arkansas*. Arkansas Geological Survey Publication GB-79-1.

Sutherland, P. K., and W. L. Manger, eds. 1979. *Mississippian-Pennsylvania Shelf-to-Basin Transition, Ozark and Ouachita Regions, Oklahoma and Arkansas*. Oklahoma Geological Survey Guidebook 19. Norman, University of Oklahoma.

Varnell, C. 2012. *Roads Less Traveled: A Journey Through the Geology and Culture of Arkansas*. Clemmons, NC: Black Jack Publishing.

8. Marlbrook Marl, Arkansas

Dane, C. H. 1929. *Upper Cretaceous Formations of Southwestern Arkansas*. Arkansas Geological Survey Bulletin 1.

Eckert, A. W. 2000. *Earth Treasures: The Southwestern Quadrant*. Lincoln, NE: iUniverse.com, Inc.

Lerman, A. 1965. "Evolution of *Exogyra* in the Late Cretaceous of the Southeastern United States." *Journal of Paleontology* 39 (3): 414–435.

Varnell, C. 2012. *Roads Less Traveled: A Journey Through the Geology and Culture of Arkansas*. Clemmons, NC: Black Jack Publishing.

9. Anza-Borrego Desert State Park, California

Jefferson, G. T. and L. Lindsay, eds. 2006. *Fossil Treasures of the Anza-Borrego Desert*. El Cajon, CA: Sunbelt Publications.

Lindsay, D., and L. Lindsay. 2006. *Anza-Borrego Desert Region: A Guide to the State Park and Adjacent Areas of the Western Colorado Desert*. Berkeley, CA: Wilderness Press.

Remeika, P., and L. Lindsay. 1992. *Geology of Anza-Borrego: Edge of Creation*. El Cajon, CA: Sunbelt Publications.

Schad, J. 1998. *Afoot and Afield in San Diego County*. Berkeley, CA: Wilderness Press.

10. Petrified Forest, California

Daniels, F. J. 1998. *Petrified Wood: The World of Fossilized Wood, Cones, Ferns, and Cycads*. Grand Junction, CO: Western Colorado Publishing.

Dorf, E. 1930. *Pliocene Floras of California*. Carnegie Institution of Washington Publication 412.

Marsh, O. C. 1871. "Notice of a Fossil Forest in the Tertiary of California." *American Journal of Science and Arts* 1 (4): 266–268.

Mattison, E. 1990. "California's Fossil Forest." *California Geology* 44 (9): 195–202.

11. La Brea Tar Pits, California

Garcia, F. A., and D. S. Miller. 1998. *Discovering Fossils: How to Find and Identify Remains of the Prehistoric Past*. Mechanicsburg, PA: Stackpole Books.

Lange, I. M. 2002. *Ice Age Mammals of North America: A Guide to the Big, the Hairy, and the Bizarre*. Missoula, MT: Mountain Press Publishing.

Meste, R. "Saber-Toothed Tales." *Discover*, April 1993. Retrieved from www.discovermagazine.com/1993/apr/sabertoothedtale202.

Reynolds, R. L. 1985. "Domestic Dog Associated with Human Remains at Rancho La Brea." *Bulletin of the Southern California Academy of Sciences* 84 (2): 76–85.

12. Sharktooth Hill, California

Barnes, L. G. 1988. *A New Fossil Pinniped (Mammalia: Otariidae) from the Middle Miocene Sharktooth Hill Bonebed, California*. Los Angeles: Natural History Museum of Los Angeles County.

Howard, H. 1966. *Additional Avian Records from the Miocene of Sharktooth Hill, California*. Los Angeles: Natural History Museum of Los Angeles County.

Mitchell, E. D. 1965. *History of Research at Sharktooth Hill, Kern County, California*. Bakersfield, CA: Kern County Historical Society.

Rintoul, W. T. 1960. "Shark Tooth Hill." *California Crossroads* 2 (5): 7–8.

13. Marsh Quarry, Colorado

Davis, B. 1998. *Bone Wars*. Wake Forest, NC: Baen Books.

Jaffe, M. 2000. *The Gilded Dinosaur: The Fossil War between E. D. Cope and O. C. Marsh and the Rise of American Science*. New York: Random House.

Kimmel. E. C. 2006. *Dinosaur Bone War: Cope and Marsh's Fossil Feud*. New York: Random House.

Lanham, U. 2011. *The Bone Hunters: The Heroic Age of Paleontology in the American West*. Mineola, NY: Dover Publications.

14. Picket Wire Track Site, Colorado

Brown, M. 1994. *Looking at Allosaurus: A Dinosaur from the Jurassic Period*. New York: Gareth Stevens Publishing.

Jenkins, J. T., and J. L. Jenkins. 1993. *Colorado's Dinosaurs*. Golden, CO: Colorado Geological Survey.

Schumacher, B. A., and M. Lockley. 2014. "Newly Documented Trackways at 'Dinosaur Lake,' the Purgatoire Valley Dinosaur Tracksite." In *Fossil Footprints of Western North America, New Mexico Museum of Natural History Bulletin* 62, edited by M. G. Lockley and S. G. Lucas, 261–267.

Williams, F., and H. Chronic. 2014. *Roadside Geology of Colorado*. 3rd ed. Missoula, MT: Mountain Press Publishing.

15. Denver Museum of Nature and Science, Colorado

Johnson, K. R. 2006. *Prehistoric Journey: A History of Life on Earth*. Golden, CO: Fulcrum Publishing.

Kramer, J., and J. Martinez. 2007. *Colorado Journey Guide: A Driving and Hiking Guide to Ruins, Rock Art, Fossils, and Formations*. Cambridge, MN: Adventure Publications.

Rudwick, M. J. S. 1985. *The Meaning of Fossils: Episodes in the History of Paleontology*. Chicago: University of Chicago Press.

Williams, F., and H. Chronic. 2014. *Roadside Geology of Colorado*. 3rd ed. Missoula, MT: Mountain Press Publishing.

16. Peabody Museum, Connecticut

Jaffe, M. 2000. *The Gilded Dinosaur: The Fossil War between E. D. Cope and O. C. Marsh and the Rise of American Science*. New York: Random House.

McCarren, M. J. 1993. *The Scientific Contributions of Othniel Charles Marsh*. New Haven, CT: Peabody Museum of Natural History.

Plate, R. 1964. *The Dinosaur Hunters: Othniel C. Marsh and Edward D. Cope*. New York: David McKay Company.

Shor, E. 1974. *The Fossil Feud between E. D. Cope and O. C. Marsh*. Detroit, MI: Exposition Press.

17. Chesapeake and Delaware Canal, Delaware

Gray, S. H. 2012. *Paleontology: The Study of Prehistoric Life*. New York: Scholastic.

Lauginiger, E. M. 1988. *Cretaceous Fossils from the Chesapeake and Delaware Canal: A Guide for Students and Collectors*. Delaware Geological Survey Special Publication 18.

Means, J. 2010. *Roadside Geology of Maryland, Delaware, and Washington, D.C.* Missoula, MT: Mountain Press Publishing.

Murray, M. 1967. *Hunting for Fossils: A Guide to Finding and Collecting Fossils in All Fifty States*. New York: Macmillan.

18. Bone Valley, Florida

Brown, R. 2013. *Florida's Fossils: A Guide to Location, Identification, and Enjoyment*. 3rd and rev. ed. Sarasota, FL: Pineapple Press.

Bryan, J. R., T. M. Scott, and G. H. Means. 2008. *Roadside Geology of Florida*. Missoula, MT: Mountain Press Publishing.

Renz, M. 1999. *Fossiling in Florida: A Guide for Diggers and Divers*. Gainesville, FL: University Press of Florida.

Ungar, P. S. 2010. *Mammal Teeth: Origin, Evolution, and Diversity*. Baltimore, MD: Johns Hopkins University Press.

19. Florida Caverns State Park, Florida

Brown, R. C. 1988. *Florida's Fossils: A Guide to Location, Identification, and Enjoyment*. Sarasota, FL: Pineapple Press.

Bryan, J. R., T. M. Scott, and G. H. Means. 2008. *Roadside Geology of Florida*. Missoula, MT: Mountain Press Publishing.

Rupert, F. R. 1994. *A Fossil Hunter's Guide to the Geology of Panhandle Florida*. Florida Geological Survey Open File Report 63.

Schmidt, W. 1988. *Florida Caverns State Park*. Florida Geological Survey Open File Report 23

20. Windley Key Fossil Reef Geological State Park, Florida

Bramson, S. H. 2011. *The Greatest Railroad Story Ever Told: Henry Flagler and the Florida East Coast Railway's Key West Extension*. Charleston, SC: History Press.

Brown, R. 2013. *Florida's Fossils: A Guide to Location, Identification, and Enjoyment*. 3rd and rev. ed. Sarasota, FL: Pineapple Press.

Harrison, R. S., and M. Coniglio. 1985. "Origin of the Pleistocene Key Largo Limestone, Florida Keys." *Bulletin of Canadian Petroleum Geology* 33 (3): 350–358.

Hoffmeister, J. E. 1974. *Land from the Sea: The Geologic Story of South Florida*. Coral Gables, FL: University of Miami Press.

21. Brevard Museum of History and Natural Science, Florida

Adovasio, J. M., R. L. Andrews, D. C. Hyland, and J. S. Illingworth. 2001. "Perishable Industries from the Windover Bog: An Unexpected Window into the Florida Archaic." *North American Archaeologist* 22 (1): 1–90.

Bryan, J. R., T. M. Scott, and G. H. Means. 2008. *Roadside Geology of Florida*. Missoula, MT: Mountain Press Publishing.

Tuross, N., M. L. Fogel, L. Newsom, and G. H. Doran. 1994. "Subsistence in the Florida Archaic: The Stable-Isotope and Archaeobotanical Evidence from the Windover Site." *American Antiquity* 59 (2): 288–303.

Wentz, R. K. 2012. *Life and Death at Windover: Excavations of a 7,000-Year-Old Pond Cemetery*. Cocoa, FL: Florida Historical Society Press.

22. Tibbs Bridge, Georgia

Fortey, R. 2001. *Trilobite: Eyewitness to Evolution*. New York: Vintage Publishing.

Gore, P. J. W., and W. Witherspoon. 2013. *Roadside Geology of Georgia*. Missoula, MT: Mountain Press Publishing.

Levi-Setti, R. 2014. *The Trilobite Book: A Visual Journey*. Chicago: University of Chicago Press.

Schaefer, L. M., 2001. *Trilobites*. Somers, NY: Richard C. Owen Publishers.

23. Makauwahi Cave, Hawaii

Blay, C., and R. Siemers. 2013. *Kauai's Geologic History: A Simplified Overview*. Koloa, HI: TEOK Investigations.

Burney, D. A. 2011. *Back to the Future in the Caves of Kaua'i: A Scientist's Adventures in the Dark*. New Haven, CT: Yale University Press.

Hazlett, R. W. 1996. *Roadside Geology of Hawaii*. Missoula, MT: Mountain Press Publishing.

Levy, S. 2008. "Lessons from a Limestone Cave: Looking to the Past to Restore the Future of a Hawaiian Island." *Wildlife Conservation*, January/February: 46–51.

24. Clarkia Fossil Bowl, Idaho

Clutter, T. 1985. "The Clarkia Fossil Bowl." *American Forests* 91 (2): 22–25.

Robertson, R. G. 1998. *Idaho Echoes in Time: Traveling Idaho's History and Geology: Stories, Directions, Maps, and More*. Mahtomedi, MN: Tamarack Books.

Smiley, C. J., and W. C. Rember. 1979. *Guidebook and Road Log to the St. Maries River (Clarkia) Fossil Area of Northern Idaho*. Idaho Bureau of Mines and Geology Information Circular 33.

Smiley, C. J., and W. C. Rember. 1985. "Physical Setting of the Miocene Clarkia Fossil Beds, Northern Idaho." In *Late Cenozoic History of the Pacific Northwest: Interdisciplinary Studies on the Clarkia Fossil Beds of Northern Idaho*, edited by C. J. Smiley, 11–31. San Francisco: Pacific Division of the American Association for the Advancement of Science.

25. Hagerman Fossil Beds National Monument, Idaho

Gidley, J. W. 1930. "A New Pliocene Horse from Idaho." *Journal of Mammalogy* 11 (3): 300–303.

MacFadden, B. J. 1994. *Fossil Horses: Systematics, Paleobiology, and Evolution of the Family Equidae*. New York: Cambridge University Press.

McDonald, H. G. 1993. "More Than Just Horses." *Rocks and Minerals* 68 (5): 322–326.

Richmond, D. R., H. G. McDonald, and J. L. Bertog. 2002. *Stratigraphy, Sedimentology, and Taphonomy of the Hagerman Horse Quarry*. Hagerman, ID: Hagerman Fossil Beds National Monument.

26. Grafton Quarries, Illinois

Fusco, G., T. Garland Jr., G. Hunt, and N. C. Hughes. 2011. "Developmental Trait Evolution in Trilobites." *Evolution* 66 (2): 314–329.

Kolata, D. R. 2011. *The Trilobite: An Early Inhabitant of Illinois*. Illinois State Geological Survey Geobit 6.

Whitney, B. R. 1969. "Trilobites of the Grafton Area (Part 1)." *Earth Science* 22 (4): 171–174.

Whitney, B. R. 1969. "Trilobites of the Grafton Area (Part 2)." *Earth Science* 22 (6): 270–274.

27. Mazonia-Braidwood State Fish and Wildlife Area, Illinois

Jennings, J. R. 1990. *Guide to Pennsylvanian Fossil Plants of Illinois*. Educational Series 13. Champaign, IL: Illinois State Geological Survey.

Nadasday, G. S. 1976. "Preliminary Investigations of the Blobs, Essex Fauna from Pit Eleven." *Proceedings of the Peoria Academy of Science* 9: 1–22.

Nitecki, M. H., ed. 1979. *Mazon Creek Fossils*. New York: Academic Press.

Shabica, C. W., and A. A. Hay, eds. 1997. *Richardson's Guide to the Fossil Fauna of Mazon Creek*. Chicago: Northeastern Illinois University Press.

28. Falls of the Ohio, Indiana

Greb, S. F., R. T. Hendricks, and D. R. Chesnut Jr. 1993. *Fossil Beds of the Falls of the Ohio*. Kentucky Geological Survey Special Publication 19, Series XI.

Rexford, C. B., and R. L Powell. 1987. "The Falls of the Ohio River, Indiana and Kentucky." In *North-Central Section of the Geological Society of America*, Centennial Field Guide, vol. 3, edited by D. L. Biggs, 381–386. Boulder, CO: Geological Society of America.

Scrutton, C. T. 1964. "Periodicity in Devonian Coral Growth." *Paleontology* 7 (part 4): 552–558.

Wells, J. W. 1963. "Coral Growth and Geochronometry." *Nature* 197 (4871): 948–950.

29. Whitewater River Gorge, Indiana

Brown, G. D., Jr., and J. A. Lineback. 1966. "Lithostratigraphy of Cincinnatian Series (Upper Ordovician) in Southeastern Indiana." *American Association of Petroleum Geologists Bulletin* 50 (5): 1018–1023.

Davis, R. A., ed. 1998. *Cincinnati Fossils: An Elementary Guide to the Ordovician Rocks and Fossils of the Cincinnati, Ohio, Region*. Cincinnati, OH: Cincinnati Museum Center.

Richards, R. P. 1972. "Autecology of Richmondian Brachiopods (Late Ordovician of Indiana and Ohio)." *Journal of Paleontology* 46 (3): 385–405.

Thompson, I. 1982. *The Audubon Society Field Guide to North American Fossils*. New York: Alfred A. Knopf.

30. Devonian Fossil Gorge, Iowa

Anderson, W. I. 1998. *Geology of Iowa: Three Billion Years of Change*. Iowa City, IA: University of Iowa Press.

Changnon, S. A. 1996. *The Great Flood of 1993: Causes, Impacts, and Responses*. Boulder, CO: Westview Press.

McGhee, G. R. 2013. *When the Invasion of Land Failed: The Legacy of the Devonian Extinctions*. New York: Columbia University Press.

Troeger, J. 1983. *From Rift to Drift: Iowa's Story in Stone*. Iowa City, IA: Iowa State Press.

31. Fossil and Prairie Park Preserve and Center, Iowa

Anderson, W. I. 1995. "The Lime Creek Formation in the Rockford Area: A Geological Perspective." In *Geology and Hydrology of Floyd-Mitchell Counties*, Geological Society of Iowa Guidebook 62, edited by B. J. Bunker, 15–20.

Day, J. 1995. "Brachiopod Fauna of the Upper Devonian (Late Frasnian) Lime Creek Formation of North-Central

Iowa and Related Units in Eastern Iowa." In *Geology and Hydrology of Floyd-Mitchell Counties*, Geological Society of Iowa Guidebook 62, edited by B. J. Bunker, 21–40.

Sorauf, J. E. 1998. *Frasnian (Upper Devonian) Rugose Corals from the Lime Creek and Shell Rock Formations of Iowa*. Bulletin of American Paleontology 113. Ithaca, NY: Paleontological Research Institution.

Wolf, R. C. 2006. *Fossils of Iowa: Field Guide to Paleozoic Deposits*. Lincoln, NE: iUniverse.com, Inc.

32. Round Mound, Kansas

Buchanan, R. 2010. *Kansas Geology: An Introduction to Landscapes, Rocks, Minerals, and Fossils*. Lawrence, KS: University Press of Kansas.

Buchanan, R., and J. R. McCauley. 2010. *Roadside Kansas: A Traveler's Guide to Its Geology and Landmarks*. Lawrence, KS: University Press of Kansas.

Mudge, M. R., and E. L. Yochelson. 1962. *Stratigraphy and Paleontology of the Uppermost Pennsylvanian and Lowermost Permian in Kansas*. US Geological Survey Professional Paper 323.

Zeller, D., ed. 1968. *The Stratigraphic Succession in Kansas*. Kansas Geological Survey Bulletin 189.

33. Sternberg Museum of Natural History, Kansas

Liggett, G. A. 2001. *Dinosaurs to Dung Beetles: Guide to the Sternberg Museum of Natural History*. Hays, KS: Sternberg Museum of Natural History.

Merriam, D. F. 1963. *The Geologic History of Kansas*. Kansas Geological Survey Bulletin 162.

Rogers, K. 1999. *The Sternberg Fossil Hunters: A Dinosaur Dynasty*. Missoula, MT: Mountain Press Publishing.

Sternberg, C. H. 1990. *The Life of a Fossil Hunter*. Bloomington, IN: Indiana University Press.

34. Big Bone Lick State Historic Site, Kentucky

Benton, M. J. 1991. *The Rise of the Mammals*. New York: Crescent Books.

Hedeen, S. 2011. *Big Bone Lick: The Cradle of American Paleontology*. Lexington, KY: University Press of Kentucky.

Jillson, W. R. 1936. *Big Bone Lick: An Outline of Its History, Geology, and Paleontology*. Louisville, KY: Standard Printing.

Jillson, W. R. 1968. *The Extinct Vertebrates of the Pleistocene in Kentucky*. Frankfort, KY: Roberts Printing.

35. Danville Bryozoan Reef, Kentucky

Karlkins, O. L. 1983. *Trepostome and Cystoporate Bryozoans from the Lexington Limestone and the Clays Ferry Formation (Middle and Upper Ordovician) of Kentucky*. US Geological Survey Professional Paper 1066-I.

Shrock, R. R., and W. H. Twenhofel. 1953. *Principles of Invertebrate Paleontology*. New York: McGraw-Hill.

Weir, G. W. 1961. *Lithostratigraphy of Upper Ordovician Strata Exposed in Kentucky*. US Geological Survey Professional Paper 1151-E.

Weir, G. W., and R. C. Greene. 1965. *Clays Ferry Formation (Ordovician): A New Map Unit in South-Central Kentucky*. US Geological Survey Bulletin 1224-B.

36. Grand Isle, Louisiana

Jones, R. W. 2014. *Foraminifera and Their Applications*. Cambridge, UK: Cambridge University Press.

Poag, C. W. 2015. *Benthic Foraminifera of the Gulf of Mexico: Distribution, Ecology, and Paleoecology*. College Station, TX: Texas A&M University Press.

Schiebout, J. A., W. A. van den Bold, and H. B. Stenzel, eds. 1986. *Montgomery Landing Site, Marine Eocene (Jackson) of Central Louisiana*. Baton Rouge, LA: Gulf Coast Association of Geological Societies.

Stringer, G. L. 2002. *46-Million-Year-Old Fossils from the Cane River Site, North-Central Louisiana*. Louisiana Geological Survey Public Information Series 10.

37. State House, Maine

Erwin, R. B. 1957. *The Geology of the Limestone of Isle La Motte and South Hero Island*. Vermont Geological Society Bulletin 9.

Norton, P. T. 1922. Fossils of the Maine State Capitol. *Maine Naturalist* 1 (4): 193–204.

Raymo, C., and M. E. Raymo. 2001. *Written in Stone: A Geological History of the Northeastern United States*. Hensonville, NY: Black Dome Press.

Welby, C. W. 1962. *Paleontology of the Champlain Basin in Vermont*. Vermont Geological Survey Special Publication 1.

38. Dinosaur Park, Maryland

Kranz, P. M. 1989. *Dinosaurs in Maryland*. Maryland Geological Survey Educational Series 6.

Kranz, P. M. 1996. "Notes on the Sedimentary Iron Ores of Maryland and Their Dinosaurian Fauna." *Maryland Geological Survey Special Publication* 3: 87–115.

Means, J. 2010. *Roadside Geology of Maryland, Delaware, and Washington, D.C.* Missoula, MT: Mountain Press Publishing.

Weishampel, D. B., and L. Young. 1996. *Dinosaurs of the East Coast*. Baltimore, MD: Johns Hopkins University Press.

39. Sandy Mile Road, Maryland

Amsden, T. W. 1951. "Paleontology of Washington County, Maryland." In *The Physical Features of Washington County, Maryland*, edited by E. Cloos, 98–123. Baltimore, MD: Maryland Department of Geology, Mines, and Water Resources.

Glaser, J. D. 1995. *Collecting Fossils in Maryland*. Maryland Geological Survey Educational Series 4.

Means, J. 2010. *Roadside Geology of Maryland, Delaware, and Washington, D.C.* Missoula, MT: Mountain Press Publishing.

Seilacher, A. 1968. "Origin and Diagenesis of the Oriskany Sandstone (Lower Devonian, Appalachians) as Reflected in Its Shell Fossils." In *Recent Developments in Carbonate Sedimentology in Central Europe*, edited by G. Müller and G. M. Friedman, 175–185. New York: Springer-Verlag.

40. Beneski Museum of Natural History, Massachusetts

Garcia, F. A., and D. S. Miller. 1998. *Discovering Fossils*. Mechanicsburg, PA: Stackpole Books.

Holtz, T. R., Jr. 2007. *Dinosaurs: The Most Complete, Up-to-Date Encyclopedia for Dinosaur Lovers of All Ages*. New York: Random House.

Hitchcock, E. 1858. *Ichnology of New England: A Report on the Sandstone of the Connecticut Valley, Especially Its Fossil Footprints*. Boston: William White, Printer to the State.

Weishampel, D. B., and L. Young. 1996. *Dinosaurs of the East Coast*. Baltimore, MD: Johns Hopkins University Press.

41. Horseshoe Harbor, Michigan

Dickas, A. B. 1986. "Comparative Precambrian Stratigraphy and Structure along the Midcontinent Rift." *American Association of Petroleum Geologists Bulletin* 70 (3): 225–238.

Elmore, R. D. 1983. "Precambrian Non-Marine Stromatolites in Alluvial Fan Deposits, the Copper Harbor Conglomerate, Upper Michigan." *Sedimentology* 30 (6): 829–842.

Planavsky, N., and M. Bjornerud. 2004. "Blowing in the Wind: The Copper Harbor Stromatolites Revisited." *Proceedings of the Institute on Lake Superior Geology* 50: 137–138.

Schopf, W. J. 2001. *Cradle of Life: The Discovery of Earth's Earliest Fossils*. Princeton, NJ: Princeton University Press.

42. Lafarge Fossil Park, Michigan

Barker, C. F. 2005. *Under Michigan: The Story of Michigan's Rocks and Fossils*. Detroit, MI: Wayne State University Press.

Dorr, J. A. 1970. *Geology of Michigan*. Ann Arbor, MI: University of Michigan Press.

Kchodl, J. J. 2006. *The Complete Guide to Michigan Fossils*. Ann Arbor, MI: University of Michigan Press.

Stack, J. R. 2014. *Fossil Hunting in the Great Lakes State: An Amateur's Guide to Fossil Hunting in Michigan*. CreateSpace Independent Publishing Platform.

43. Petoskey State Park, Michigan

Gardner, W. C. 1974. "Middle Devonian Stratigraphy and Depositional Environments in the Michigan Basin." *Michigan Basin Geological Society Special Papers* 1: 43–47.

Kesling, R. V., R. T. Segall, and H. O. Sorensen. 1974. *Devonian Strata of Emmet and Charlevoix Counties, Michigan*. Papers on Paleontology 7. Ann Arbor, MI: University of Michigan Museum of Paleontology.

Milstein, R. L. 1987. "Middle Devonian Transverse Group in Charlevoix and Emmet Counties, Michigan." In *North-Central Section of the Geological Society of America, Centennial Field Guide*, vol. 3, edited by D. L. Biggs, 293–296. Boulder, CO: Geological Society of America.

Mueller, B., and W. H. Wilde. 2004. *The Complete Guide to Petoskey Stones*. Ann Arbor, MI: University of Michigan Press.

44. Gunflint Chert, Minnesota

Barghoorn, E. S., and S. A. Tyler. 1965. "Microorganisms from the Gunflint Chert." *Science* 147 (3658): 563–575.

Darwin, C. *On the Origin of Species*. London: John Murray, Albemarle Street.

Hussey, R. C. 1947. *Historical Geology*. New York: McGraw Hill.

Schopf, J. W. 1999. *Cradle of Life: The Discovery of Earth's Earliest Fossils*. Princeton, NJ: Princeton University Press.

Tyler, S., and E. S. Barghoorn. 1954. "Occurrence of Structurally Preserved Plants in Pre-Cambrian Rocks of the Canadian Shield." *Science* 119 (3096): 606–608.

45. Wangs, Minnesota

Balaban, N. H., ed. 1987. *Field Trip Guidebook for the Upper Mississippi Valley, Minnesota, Iowa, and Wisconsin*. Minnesota Geological Survey Guidebook 15.

Mossler, J., and S. Benson. 2006. *Minnesota at a Glance: Fossil Collecting in the Twin Cities Area*. St. Paul, MN: Minnesota Geological Survey, University of Minnesota.

Ojakangas, R. W., and C. L. Matsch. 1982. *Minnesota's Geology*. Minneapolis, MN: University of Minnesota Press.

Hogberg, R. K., R. E. Sloan, and S. Tufford. 1965. *Guide to Fossil Collecting in Minnesota*. Minnesota Geological Survey Educational Series 1.

46. W. M. Browning Cretaceous Fossil Park, Mississippi

Baldwin, M. 2003. "Frankstown Paleoecology." *Rockhound News* 49 (8): 1–8.

Cocke, J. 2002. *Fossil Shark Teeth of the World*. Redondo Beach, CA: Lamna Books.

Dockery, D. T., and D. E. Thompson. 2016. *The Geology of Mississippi*. Jackson, MS: University Press of Mississippi.

Manning, E. M., and D. T. Dockery, III. 1992. *A Guide to the Frankstown Vertebrate Fossil Locality (Upper Cretaceous), Prentiss County, Mississippi*. Mississippi Department of Environmental Quality, Office of Geology Circular 4.

47. Mastodon State Historic Site, Missouri

Chapman, C. 1980. *The Archeology of Missouri*. Vol. 2. Columbia, MO: University of Missouri Press.

Haynes, G. 2002. *The Early Settlement of North America: The Clovis Culture*. New York: Cambridge University Press.

Spencer, C. G. 2011. *Roadside Geology of Missouri*. Missoula, MT: Mountain Press Publishing.

Thomson, K. S. 2008. *The Legacy of the Mastodon: The Golden Age of Fossils in America*. New Haven, CT: Yale University Press.

48. Riverbluff Cave, Missouri

Brown, G. 1993. *Great Bear Almanac*. Lanham, MD: Lyons Press.

Spencer, C. G. 2011. *Roadside Geology of Missouri*. Missoula, MT: Mountain Press Publishing.

Unklesbay, A. G. 1955. *The Common Fossils of Missouri*. Columbia, MO: University of Missouri Press.

Weaver, D. H. 2008. *Missouri Caves in History and Legend*. Columbia, MO: University of Missouri Press.

49. Hell Creek Country, Montana

Alt, D., and D. W. Hyndman. 1986. *Roadside Geology of Montana*. Missoula, MT: Mountain Press Publishing.

Bird, R. T., ed. 1985. *Bones for Barnum Brown*. Fort Worth, TX: Texas Christian University Press.

Dingus, L. 2015. *Hell Creek, Montana: America's Key to the Prehistoric Past*. New York: St. Martin's Press.

Hartman, J. H. 2014. *Through the End of the Cretaceous in the Type Locality of the Hell Creek Formation in Montana and Adjacent Areas*. Special Paper 503. Boulder, CO: Geological Society of America.

50. Museum of the Rockies, Montana

Cadbury, D. 2001. *Terrible Lizard: The First Dinosaur Hunters and the Birth of a New Science*. New York: Henry Holt.

Dingus, L., and M. Norell. 2010. *Barnum Brown: The Man Who Discovered* Tyrannosaurus rex. Oakland, CA: University of California Press.

Jaffe, M. 2000. *The Gilded Dinosaur: The Fossil War between E. D. Cope and O. C. Marsh and the Rise of American Science*. New York: Random House.

Powell, J. L. 1998. *Night Comes to the Cretaceous: Dinosaur Extinction and the Transformation of Modern Geology*. New York: W. H. Freeman.

51. Agate Fossil Beds National Monument, Nebraska

Holmes, T. 2008. *The Age of Mammals: The Oligocene and Miocene Epochs*. New York: Chelsea House Publishing.

Johnson, K. 2007. *Cruisin' the Fossil Freeway*. Golden, CO: Fulcrum Publishing.

Maher, H. D., Jr., G. F. Engelmann, and R. D. Schuster. 2003. *Roadside Geology of Nebraska*. Missoula, MT: Mountain Press Publishing.

Prothero, D. R. 2006. *After the Dinosaurs: The Age of Mammals*. Bloomington, IN: Indiana University Press.

52. Ashfall Fossil Beds State Historical Park, Nebraska

Bonnichsen, B. 1982. "The Bruneau-Jarbidge Eruptive Center, Southwestern Idaho." In *Cenozoic Geology of Idaho*, Idaho Bureau of Mines and Geology Bulletin 26, edited by B. Bonnichsen and R. M. Breckenridge, 217–254.

Rogers, R. R., D. A. Eberth, and A. R. Fiorillo eds. 2008. *Bonebeds: Genesis, Analysis, and Paleobiological Significance*. Chicago: University of Chicago Press.

Rose, W. I., C. M. Riley, and S. Dartevelle. 2003. "Sizes and Shapes of 10-Ma Distal Fall Pyroclasts in the Ogallala Group, Nebraska." *Journal of Geology* 111 (1): 115–124.

Voorhies, M. 1992. *Ashfall: Life and Death at a Nebraska Waterhole Ten Million Years Ago*. University of Nebraska State Museum, Museum Notes 81. Retrieved from http://ashfall.unl.edu/life_death.html.

53. Little Blue Outcrop, Nebraska

Everhart, M. J. 2005. *Oceans of Kansas: A Natural History of the Western Interior Seaway*. Bloomington: Indiana University Press.

Hattin, D. E. 1975. *Stratigraphy and Depositional Environment of the Greenhorn Limestone (Upper Cretaceous) of Kansas*. Kansas Geological Survey Bulletin 209.

Maher, H. D., Jr., G. F. Engelmann, and R. D. Schuster. 2003. *Roadside Geology of Nebraska*. Missoula, MT: Mountain Press Publishing.

Pabian, R. K. 1970. *Record in Rock: A Handbook of the Invertebrate Fossils of Nebraska*. Educational Circular 1. Lincoln, NE: University of Nebraska.

54. Berlin-Ichthyosaur State Park, Nevada

Camp, C. L. 1981. *Child of the Rocks: The Story of Berlin-Ichthyosaur State Park*. Nevada Bureau of Mines and Geology Special Publication 5.

Kosch, B. F. 1990. "A Revision of the Skeletal Reconstruction of *Shonisaurus popularis*." *Journal of Vertebrate Paleontology* 19 (1): 42–49.

Orndorff, R. L., R. W. Wieder, and H. F. Filkorn. 2000. *Geology Underfoot in Central Nevada*. Missoula, MT: Mountain Press Publishing.

Wallace, D. R. 2007. *Neptune's Ark: From Ichthyosaurs to Orcas*. Oakland, CA: University of California Press.

55. Virgin Valley, Nevada

Helymun, E. B. 1987. "Virgin Valley." *Lapidary Journal* 41 (3): 33–44.

Leechman, F. 1982. *The Opal Book: A Complete Guide to the Famous Gem*. Naremburn, North South Wales, Australia: Lansdowne Press.

Ward, F. 2011. *Opals*. Malibu, CA: Gem Book Publishers.

Zeitner, J. C. 1986. "Precious Opal in the United States." *Lapidary Journal* 40 (3): 42–48.

56. Odiorne Point State Park, New Hampshire

Harrison, W., and C. J. Lyon. 1963. "Sea-Level and Crustal Movements along the New England–Acadian Shore, 4,500–3,000 B.P." *Journal of Geology* 71 (1): 96–108.

Lyon, C. J., and J. W. Goldthwait. 1934. "An Attempt to Cross-Date Trees in Drowned Forests." *Geographical Review* 24 (4): 605–614.

Raymo, C., and M. E. Raymo. 2001. *Written in Stone: A Geological History of the Northeastern United States*. Hensonville, NY: Black Dome Press.

Van Diver, B. B. 1987. *Roadside Geology of Vermont and New Hampshire*. Missoula, MT: Mountain Press Publishing.

57. Big Brook Preserve, New Jersey

Cocke, J. 2002. *Fossil Shark Teeth of the World: A Collector's Guide*. Torrance, CA: Lamna Books Publishing.

Ferrari, A., and A. Ferrari. 2002. *Sharks*. Buffalo, NY: Firefly Books.

Kent, B. W. 1994. *Fossil Sharks of the Chesapeake Bay Region.* Columbia, MD: Egan Rees and Boyer.

Rathbone, J. 2012. *Pictorial Guide to Fossil Shark Teeth from Around the World.* CreateSpace Independent Publishing Platform.

58. Hamburg Stromatolites, New Jersey

Allwood, A. C., M. R. Walter, B. S. Kamber, C. P. Marshall, and I. W. Burch. 2006. "Stromatolite Reef from the Early Archean Era of Australia.*" Nature* 441 (7094): 714–718.

Grotzinger, J. P., and A. H. Knoll. 1999. "Stromatolites in Precambrian Carbonates: Evolutionary Mileposts or Environmental Dipsticks?" *Annual Review of Earth and Planetary Sciences* 27: 313–358.

Logan, B. W., R. Rezak, and R. N. Ginsbury. 1964. "Classification and Environmental Significance of Algal Stromatolites." *Journal of Geology* 72 (1): 68–83.

Schopf, J. W. 2001. *Cradle of Life: The Discovery of Earth's Earliest Fossils.* Princeton, NJ: Princeton University Press.

59. Battleship Rock, New Mexico

Kues, B. S. 1982. *Fossils of New Mexico.* Albuquerque, NM: University of New Mexico Press.

Kues, B. S. 1996. "Guide to the Late Pennsylvanian Paleontology of the Upper Madera Formation, Jemez Springs, North-Central New Mexico." In *Jemez Mountains Region*, New Mexico Geological Society 47th Annual Fall Field Conference Guidebook, edited by F. Goff, B. S Kues, M. A. Rogers, L. S. McFadden, and L. S. Gardner, 169–188. Socorro, NM: New Mexico Geological Society.

Kues, B. S. 2008. *The Paleontology of New Mexico.* Albuquerque, NM: University of New Mexico Press.

Voynick, S. M. 1997. *New Mexico: A Guide to Minerals, Gemstones, and Fossils.* Missoula, MT: Mountain Press Publishing.

60. Ghost Ranch, New Mexico

Ahmed, B. C. A., and G. Kocurek. 2000. "Catastrophic Flooding of an Aeolian Dune Field: Jurassic Entrada and Todilto Formations, Ghost Ranch, New Mexico, USA." *Sedimentology* 47 (6): 1069–1080.

Colbert, E. H. 1995. *The Little Dinosaurs of Ghost Ranch.* New York: Columbia University Press.

Heckert, A. B., S. G. Lucas, A. Downs, and A. P. Hunt. 2005. "Fossils at Ghost Ranch: The Ruth Hall Museum of Paleontology Collection." In *Geology of Chama Basin*, New Mexico Geological Society Guidebook 56, edited by S. G. Lucas, K. E. Zeigler, V. W. Lueth, and D. E. Owen, 40–41. Socorro, NM: New Mexico Geological Society.

Nesbitt, S. J., M. R. Stocker, B. J. Small, and A. Downs. 2009. "The Osteology and Relationships of *Vancleavea campi* (Reptilia: Archosauriformes)." *Zoological Journal of the Linnean Society* 157 (4): 814–864.

61. Clayton Lake State Park, New Mexico

Delgalvis, A. 2015. *The Lost Tracks: A Journey of Discovery.* Grand Junction, CO: Studio 2138 LLC.

Falkingham, P. L., D. Marty, and A. Richter, eds. 2016. *Dinosaur Tracks: The Next Steps.* Bloomington, IN: Indiana University Press.

Lessem, D. 2010. *National Geographic Kids Ultimate Dinopedia: The Most Complete Dinosaur Reference Ever.* Washington DC: National Geographic.

Lockley, M., and A. P. Hunt. 1999. *Dinosaur Tracks of Western North America.* New York: Columbia University Press.

62. John Boyd Thacher State Park, New York

Goldring, W. 1997. *Guide to the Geology of John Boyd Thacher Park (Indian Ladder Region) and Vicinity.* New York State Museum Handbook 14. Albany, NY: University of the State of New York, State Education Dept., New York State Museum, Cultural Education Center.

Laporte, L. F. 1967. "Carbonate Deposition near Mean Sea-Level and Resultant Facies Mosaic Manlius Formation (Lower Devonian) of New York State." *American Association of Petroleum Geologists Bulletin* 51 (1): 73–101.

Van Diver, B. B. 1985. *Roadside Geology of New York.* Missoula, MT: Mountain Press Publishing.

Ver Straeten, C. 2007. "Rock of Deep Ages: The Indian Ladder Trail, Ancient Thacher Park." *Legacy: The Magazine of the New York State Museum* 3 (1): 10–11.

63. Penn Dixie Fossil Park and Nature Reserve, New York

Collins, C. 1995. *Care and Conservation of Paleontological Material.* Amsterdam, Netherlands: Elsevier Publishing.

Landing, E. 2004. *Fossils and "Deep Time" in New York*. New York: New York State Museum.

Whiteley, T. E. 2002. *Trilobites of New York: An Illustrated Guide*. Barcelona, Spain: Comstock Publishing Associates.

Wilson, K. A. 2014. *Field Guide to the Devonian Fossils of New York*. Special Publication 44. Ithaca, NY: Paleontological Research Institution.

64. Aurora Fossil Museum, North Carolina

Bretton, W. 1994. *Fossil Sharks of the Chesapeake Bay Region*. Columbia, MD: Egan Rees and Boyer.

Cocke, J. 2002. *Fossil Shark Teeth of the World*. Redondo Beach, CA: Lamna Books.

Oliphant, A. 2015. *Shark Teeth Hunting on the Carolina Coast*. Sarasota, FL: Pineapple Press.

Renz, M. 2002. *Megalodon: Hunting the Hunter*. Walla Walla, WA: Paleo Press.

65. Green Mill Run, North Carolina

Civard-Racinais, A., and P. Héraud. 2012. *Great White Shark: Myth and Reality*. Richmond Hill, Ontario: Firefly Books.

Cocke, J. 2002. *Fossil Shark Teeth of the World*. Redondo Beach, CA: Lamna Books.

Ellis, R., and J. McCosker. 1995. *Great White Shark*. Redwood City, CA: Stanford University Press.

Oliphant, A. 2015. *Shark Tooth Hunting on the Carolina Coast*. Sarasota, FL: Pineapple Press.

66. Heritage Center and State Museum, North Dakota

Everhart, M. J. 2005. *Oceans of Kansas: A Natural History of the Western Interior Sea*. Bloomington, IN: Indiana University Press.

Haines, T., and P. Chambers. 2006. *The Complete Guide to Prehistoric Life*. Buffalo, NY: Firefly Books.

Parrish, J. M., R. E. Molnar, P. J. Currie, and E. B. Koppelhus, eds. 2013. *Tyrannosaurid Paleobiology*. Bloomington, IN: Indiana University Press.

Powell, J. L. 1998. *Night Comes to the Cretaceous: Dinosaur Extinction and the Transformation of Modern Geology*. New York: W. H. Freeman.

67. Pembina Gorge, North Dakota

Ellis, R. 2003. *Sea Dragons: Predators of the Prehistoric Oceans*. Lawrence, KS: University Press of Kansas.

Gill, J. R., and W. A. Cobban. 1965. *Stratigraphy of the Pierre Shale, Valley City and Pembina Mountain Areas, North Dakota*. US Geological Survey Professional Paper 392.

Stinchcomb, B. L. 2014. *Mesozoic Fossils II: The Cretaceous Period*. Atglen, PA: Schiffer Publishing.

Trimble, D. E. 2000. *The Geologic Story of the Great Plains*. Medora, ND: Theodore Roosevelt Nature and History Association.

68. Caesar Creek State Park, Ohio

Davis, R. A., ed. 1998. *Cincinnati Fossils: An Elementary Guide to the Ordovician Rocks and Fossils of the Cincinnati, Ohio, Region*. Cincinnati, OH: Cincinnati Museum of Natural History.

Feldmann, R. M., and M. Hackathorn, eds. 1996. *Fossils of Ohio*. Ohio Division of Geological Survey Bulletin 70.

Shrake, D. L. 1992. *Excursion to Caesar Creek State Park in Warren County, Ohio: A Classic Upper Ordovician Fossil-Collecting Locality*. Ohio Division of Geological Survey Guidebook 12.

Shrake, D. L. 1994. *Ohio Trilobites*. Ohio Division of Geological Survey GeoFacts 5.

69. Sylvania Fossil Park, Ohio

Camp, M. J., and C. B. Hatfield. 1991. "Middle Devonian (Gevetian) Silica Formation of Northwest Ohio: Description and Road Log." *Ohio Journal of Science* 91 (1): 27–34.

Hallam, A., and P. B. Wignall. 1997. *Mass Extinctions and Their Aftermath*. Oxford, UK: Oxford University Press.

McGhee, G. R., Jr. 1996. *The Late Devonian Mass Extinction: The Frasnian/Famennian Crisis*. New York: Columbia University Press.

Stewart, G. A. 1927. *Fauna of the Silica Shale of Lucas County*. Ohio Division of Geological Survey Bulletin 32.

70. Arbuckle Anticline, Oklahoma

Copper, P. 1996. *Brachiopods*. Boca Raton, FL: CRC Press.

Fay, R. O. 1988. "I-35 Roadcuts: Geology of Paleozoic Strata in the Arbuckle Mountains of Southern Oklahoma." In *South-Central Section of the Geological Society of America*,

Centennial Field Guide, vol. 4, edited by O. T. Howard, 183–188. Boulder, CO: Geological Society of America.

Fay, R. O. 1989. *Geology of the Arbuckle Mountains along Interstate 35, Carter and Murray Counties, Oklahoma*. Oklahoma Geological Survey Guidebook 26. Norman, OK: University of Oklahoma Press.

Ham, W. E., and J. H. Stitt. 1969. *Regional Geology of the Arbuckle Mountains, Oklahoma*. Oklahoma Geological Survey Guidebook 17. Norman, OK: University of Oklahoma Press.

71. John Day Fossil Beds National Monument, Oregon

Graham, J. P. 2014. *John Day Fossil Beds National Monument: Geologic Resources Inventory Report*. Fort Collins, CO: National Park Service.

Justice, K. 2012. *Explore John Day Fossil Beds with Noah Justice*. Green Forest, AR: Master Books.

Lanham, U. 2011. *The Bone Hunters: The Heroic Age of Paleontology in the American West*. Mineola, NY: Dover. Publications.

Sternberg, C. H. 1990. *The Life of a Fossil Hunter*. Bloomington, IN: Indiana University Press.

72. Lava Cast Forest, Oregon

Harris, S. L. 2005. *Fire Mountains of the West: The Cascade and Mono Lake Volcanoes*. Missoula, MT: Mountain Press Publishing.

MacLeod, N. S., and D. R. Sherrod. 1981. "Newberry Volcano, Oregon." In *Guides to Some Volcanic Terranes in Washington, Idaho, Oregon, and Northern California*, US Geological Survey Circular 838, edited by D. A. Johnston and J. Donnelly-Nolan, 85–92.

Miller, M. B. 2014. *Roadside Geology of Oregon*. Missoula, MT: Mountain Press Publishing.

Sherrod, D. R., L. G. Mastin, W. E. Scott, and S. P. Schilling. 1997. *Volcano Hazards at Newberry Volcano, Oregon*. US Geological Survey Open-File Report 97-513.

73. Montour Preserve, Pennsylvania

Garcia, F. A. 1998. *Discovering Fossils: How to Find and Identify Remains of the Prehistoric Past*. Mechanicsburg, PA: Stackpole Books.

McGhee, G. R., Jr. 1996. *The Late Devonian Mass Extinction: The Frasnian/Famennian Crisis*. New York: Columbia University Press.

McGhee, G. R., Jr. 2013. *When the Invasion of Land Failed: The Legacy of the Devonian Extinctions*. New York: Columbia University Press.

Wilson, K. A. 2014. *Field Guide to the Devonian Fossils of New York*. Special Publication 44. Ithaca, NY: Paleontological Research Institution.

74. Red Hill, Pennsylvania

Clack, J. 2002. *Gaining Ground: The Origin and Evolution of the Tetrapods*. Bloomington, IN: Indiana University Press.

Daeschler, E. B. 2000. "Early Tetrapod Jaws from the Late Devonian of Pennsylvania, U.S.A." *Journal of Paleontology* 74 (2): 301–308.

Daeschler, E. B., and N. H. Shubin. 1998. "Fish with Fingers?" *Nature* 391: 133.

Zimmer, C. 1998. *At the Water's Edge: Macroevolution and the Transformation of Life*. New York: Free Press.

75. Cory's Lane, Rhode Island

Jones, C. 2006. *The History and Future of Narragansett Bay*. Boca Raton, FL: Universal Publishers.

Lyons, P. C., and W. C. Darrah. 1978. "A Late Middle Pennsylvanian Flora of the Narragansett Basin, Massachusetts." *Geological Society of America Bulletin* 89 (3): 433–438.

Skehan, J. W. 2008. *Roadside Geology of Connecticut and Rhode Island*. Missoula, MT: Mountain Press Publishing.

Stinchcomb, B. L. 2013. *Paleozoic Fossil Plants*. Atglen, PA: Schiffer Publishing.

76. Edisto Beach, South Carolina

Ladd, H. S. 1939. "Land Animals from the Sea." *Regional Review* 3 (3). Retrieved from https://www.nps.gov/parkhistory/online_books/regional_review/vol3-3b.htm.

Roth, J. A., and J. Laerm. 1980. "A Late Pleistocene Vertebrate Assemblage from Edisto Island, South Carolina." *Brimleyana* 3: 1–29.

Sanders, A. E. 2002. *Additions to the Pleistocene Mammal Faunas of South Carolina, North Carolina, and Georgia*. Transactions of the American Philosophical Society, vol. 92, pt. 5. Philadelphia: American Philosophical Society.

Witherington, B., and D. Witherington. 2011. *Seashells of Georgia and the Carolinas*. Sarasota, FL: Pineapple Press.

77. Mammoth Site, South Dakota

Gries, J. P. 1996. *Roadside Geology of South Dakota*. Missoula, MT: Mountain Press Publishing.

Lister, A., and P. G. Bahn. 2007. *Mammoths: Giants of the Ice Age*. London: Francis Lincoln Publishers.

Martin, P. S. 2007. *Twilight of the Mammoths: Ice Age Extinctions and the Rewilding of America*. Berkeley, CA: University of California Press.

Nelson, L. W. 2005. *Mammoth Graveyard: A Treasure Trove of Clues to the Past*. Rapid City, SD: Feånwyn Press.

78. Petrified Wood Park, South Dakota

Daniels, F. J. 1998. *Petrified Wood: The World of Fossilized Wood, Cones, Ferns, and Cycads*. Grand Junction, CO: Western Colorado Publishing.

Daniels, F. J., and R. D. Dayvault. 2006. *Ancient Forests: A Closer Look at Fossil Wood*. Grand Junction, CO: Western Colorado Publishing.

Gries, J. P. 1996. *Roadside Geology of South Dakota*. Missoula, MT: Mountain Press Publishing.

Hickey, L. J. 2010. *The Forest Primeval: The Geologic History of Wood and Petrified Forests*. New Haven, CT: Yale Peabody Museum.

79. Coon Creek Science Center, Tennessee

Adams, S., and K. Adams. 1994. "Creatures in the Coon Creek Clay." *Tennessee Conservationist* LX (1): 8–13.

Barnes, B. 1989. "A Tennessee Geological Treasure Chest." *Tennessee Magazine* 33 (4): 8–15.

Eckert, A. W. 1963. "Coon Creek's Fabulous Fossils." *Science Digest* 52 (1): 47–53.

Ehret, D., L. Harrell, and S. Ebersole, eds. 2016. *Paleontology of Cretaceous Coon Creek*. Alabama Museum of Natural History Bulletin 33. Tuscaloosa, AL: University of Alabama.

80. Gray Fossil Site, Tennessee

Kohl, M. S. 2001. *Subsidence and Sinkholes in East Tennessee: A Field Guide to Holes in the Ground*. Tennessee Division of Geology Public Information Series 1.

Miller, R. A. 2008. *The Geologic History of Tennessee*. Tennessee Division of Geology Bulletin 74.

Moore, H. 2004. *The Bone Hunters: The Discovery of Miocene Fossils in Gray, Tennessee*. Knoxville, TN: University of Tennessee Press.

White, W. B. 1988. *Geomorphology and Hydrology of Karst Terrains*. New York: Oxford University Press.

81. Rock Island State Park, Tennessee

Hess, H., W. I. Ausich, C. E. Brett, and M. J. Simms 1999. *Fossil Crinoids*. New York: Cambridge University Press.

Lane, G. N., and W. I. Ausich. 2001. "The Legend of St. Cuthbert's Beads: A Palaeontological and Geological Perspective." *Folklore* 112 (1): 65–73.

Morgan, W. W. 2014. *Collector's Guide to Crawfordsville Crinoids*. Atglen, PA: Schiffer Publishing.

Rhodes, F. H. T., H. S. Zim, and P. R. Shaffer. 2001. *Fossils: A Guide to Prehistoric Life*. New York: St. Martin's Press.

82. Fluvanna Roadcut, Texas

Finsley, C. E. 1999. *A Field Guide to Fossils of Texas*. New York: Gulf Publishing.

Matthews, W. H. 1988. *Texas Fossils: An Amateur Collector's Handbook*. Austin, TX: University of Texas, Bureau of Economic Geology.

Morris, J. M. 2013. *El Llano Estacado: Explorations and Imagination on the High Plains of Texas and New Mexico, 1536–1860*. Austin, TX: Texas State Historical Association.

Temple, R. D. 2008. *Edge Effects: The Border-Name Places*. Lincoln, NE: iUniverse.com, Inc.

83. Ladonia Fossil Park, Texas

Finsley, C. E. 1999. *A Field Guide to Fossils of Texas*. New York: Gulf Publishing.

Matthews, W. H. 1988. *Texas Fossils: An Amateur Collector's Handbook*. Austin, TX: University of Texas, Bureau of Economic Geology.

McKinzie, M. G., R. Morin, and E. Swiatovy. 2001. *Fossil Collector's Guide to the North Sulphur River*. Dallas, TX: Dallas Paleontological Society.

Spearing, D. 1991. *Roadside Geology of Texas*. Missoula, MT: Mountain Press Publishing.

84. Waco Mammoth National Monument, Texas

Haynes, G. 1993. *Mammoths, Mastodons, and Elephants: Biology, Behavior, and the Fossil Record*. Cambridge, UK: Cambridge University Press.

Lange, I. M. 2002. *Ice Age Mammals of North America: A Guide to the Big, the Hairy, and the Bizarre*. Missoula, MT: Mountain Press Publishing.

Lister, A. 2014. *Mammoths and Mastodons of the Ice Age*. Richmond Hill, Ontario: Firefly Books.

Lister, A., and P. Bahn. 1994. *Mammoths*. New York: MacMillan Press.

85. Mineral Wells Fossil Park, Texas

Finsley, C. E. 1999. *A Field Guide to Fossils of Texas*. New York: Gulf Publishing.

Matthews, W. H. 1988. *Texas Fossils: An Amateur Collector's Handbook*. Austin, TX: University of Texas, Bureau of Economic Geology.

McKinzie, M., and J. McLeod. 2012. *Pennsylvania Fossils of North Texas*. Occasional Papers of the Dallas Paleontological Society, vol. 6.

Swanson, E. R. 1996. *Geo-Texas: A Guide to the Earth Sciences*. College Station, TX: Texas A&M University Press.

86. U-Dig Fossils Quarry, Utah

Fortey, R. 2001. *Trilobite: Eyewitness to Evolution*. New York: Vintage Publishing.

Green, J. 2007. *A Monograph of the Trilobites of North America: With Colored Models of the Species*. Kila, MT: Kessinger Publishing.

Lawrence, P., and S. Stammers. 2014. *Trilobites of the World: An Atlas of 1,000 Photographs*. Rochdale, UK: Siri Scientific Press.

Levi-Setti, R. 2014. *The Trilobite Book: A Visual Journey*. Chicago: University of Chicago Press.

87. Wall of Bones, Utah

Dixon, D. 2008. *World Encyclopedia of Dinosaurs and Prehistoric Creatures*. London: Anness Publishing.

Foster, J. 2007. *Jurassic West: The Dinosaurs of the Morrison Formation and Their World*. Bloomington, IN: Indiana University Press.

Holtz, T. R., Jr., and M. Brett-Surman. 2015. *Jurassic World Dinosaur Field Guide*. New York: Random House Books for Young Readers.

Williams, F., L. Chronic, and H. Chronic. 2014. *Roadside Geology of Utah*. 2nd ed. Missoula, MT: Mountain Press Publishing.

88. Chazy Fossil Reef National Natural Landmark, Vermont

James, N. P., and P. Bourque. 1992. "Reefs and Mounds." In *Facies Models: Response to Sea Level Change*, edited by R. G. Walker and N. P. James, 332–348. St. John's, Newfoundland: Geological Association of Canada.

Pitcher, M. 1964. "Evolution of Chazyan (Ordovician) Reefs of Eastern United States and Canada." *Bulletin of Canadian Petroleum Geology* 12 (3): 632–691.

Stock, C. W. 2001. "Stromatoporoidea, 1926–2000." *Journal of Paleontology* 75 (6): 1079–1089.

Teresi, D. 2007. "What's the World's Oldest Communal Ocean Reef Doing in the Green Mountain State?" *Smithsonian Magazine*, January. Retrieved from http://www.smithsonianmag.com/science-nature/paleozoic-vermont-143627957.

89. Chippokes Plantation State Park, Virginia

Burns, J. 1991. *Fossil Collecting in the Mid-Atlantic States: With Localities, Collecting Tips, and Illustrations of More Than 450 Fossil Specimens*. Baltimore, MD: Johns Hopkins University Press.

Frye, K. 1986. *Roadside Geology of Virginia*. Missoula, MT: Mountain Press Publishing.

Say, T. 1824. *An Account of Some of the Fossil Shells of Maryland*. Philadelphia: Academy of Natural Sciences of Philadelphia.

Ward, L., and B. Blackwelder. 1975. *Chesapecten, a New Genus of Pectinidae (Mollusca: Bivalvia) from the Miocene and Pliocene of Eastern North America*. US Geological Survey Professional Paper 861.

90. Hogue Creek, Virginia

Parker, P. M., ed. 2009. *Brachiopods: Webster's Timeline History, 1809–2007*. Las Vegas, NV: ICON Group International.

Rudwick, M. J S. 1979. *Living and Fossil Brachiopods*. London: Humanities Press.

Tillman, J. R. 1964. "Variation in Species of *Mucrospirifer* from Middle Devonian Rocks of Michigan, Ontario, and Ohio." *Journal of Paleontology* 38 (5): 952–964.

Walker, C., and D. Ward. 2002. *Fossils*. Smithsonian Handbooks. London: Dorling Kindersley Limited.

91. Mint Spring, Virginia

Fortey, R. 1998. *Life: A Natural History of the First Four Billion Years of Life on Earth*. New York: Alfred A. Knopf.

Palmer, D., and B. Rickards, eds. 2001. *Graptolites: Writing in the Rocks*. Rochester, NY: Boydell and Brewer.

Ruppert, E. E., and R. D. Barnes. 1994. *Invertebrate Zoology*. Fort Worth, TX: Saunders College Publishing.

Svitil, K. A. 1993. "It's Alive, and It's a Graptolite." *Discover Magazine*, July: 18–19.

92. Ginkgo Petrified Forest State Park, Washington

Brockman, C. F. 2012. *The Story of the Petrified Forest: Ginkgo State Park, Washington*. Whitefish, MT: Literary Licensing.

Daniels, F. J. 1998. *Petrified Wood: The World of Fossilized Wood, Cones, Ferns, and Cycads*. Grand Junction, CO: Western Colorado Publishing.

Daniels, F. J., and R. D. Dayvault. 2006. *Ancient Forests: A Closer Look at Fossil Wood*. Grand Junction, CO: Western Colorado Publishing.

Hickey, L. J. 2010. *The Forest Primeval: The Geologic History of Wood and Petrified Forests*. New Haven, CT: Yale Peabody Museum.

93. Stonerose Interpretive Center and Eocene Fossil Site, Washington

Alt, D. D., and D. Hyndman. 1984. *Roadside Geology of Washington*. Missoula, MT: Mountain Press Publishing.

Mustoe, G. E. 2015. "Geologic History of Eocene Stonerose Fossil Beds, Republic, Washington, USA." *Geosciences* 5 (3): 243–263.

Wehr, W. C., and D. Q. Hopkins. 1994. "The Eocene Orchards and Gardens of Republic, Washington." *Washington Geology* 22 (3): 27–34.

Wolfe, J. A., and W. C. Wehr. 1991. "Significance of the Eocene Fossil Plants at Republic." *Washington Geology* 19 (3): 18–24.

94. Glen Lyn Roadcut, West Virginia

Gray, S. H. 2005. *Crinoids and Blastoids*. North Mankato, MN: Child's World.

Sundberg, F. A. 1990. "Upper Carboniferous (Namurian) Amphibian Trackways from the Bluefield Formation, West Virginia, USA." *Ichnos* 1 (2): 111–124.

Whisonant, R. C., and A. P. Schultz. 1986. "Appalachian Valley and Ridge to Appalachian Plateau Transition Zone in Southwestern Virginia and Eastern West Virginia: Structure and Sedimentology." In *Southeastern Section of the Geological Society of America*, Centennial Field Guide, vol. 6, edited by T. L. Neathery, 113–118. Boulder, CO: Geological Society of America.

Whisonant, R. C., and R. J. Scolaro. 1980. "Tide-Dominated Coastal Environments in the Mississippian Bluefield Formation of Eastern West Virginia." In *Shorelines, Past and Present*, Proceedings of the Fifth Symposium on Coastal Sedimentology, 25–26 January, edited by W. F. Tanner, 593–637. Tallahassee, FL: Department of Geology, Florida State University.

95. Lost River Quarry, West Virginia

Fortey, R. 2001. *Trilobite: Eyewitness to Evolution*. New York: Vintage Publishing.

Levi-Setti, R. 1995. *Trilobites*. Chicago: University of Chicago Press.

Mikulic, D. G. 2007. *Fabulous Fossils: 300 Years of Worldwide Research on Trilobites*. Albany, NY: New York State Museum.

Schaefer, L. M., 2001. *Trilobites*. Somers, NY: Richard C. Owen Publishers.

96. Blackberry Hill, Wisconsin

Collette, J. H., J. W. Hagadorn, and M. A. Lacelle. 2010. "Dead in Their Tracks: Cambrian Arthropods and Their Traces from Intertidal Sandstones of Quebec and Wisconsin." *Palaios* 25 (8): 475–486.

Getty, P. R., and J. W. Hagadorn. 2009. "Paleobiology of the *Climactichnites* tracemaker." *Paleontology* 52 (4): 753–778.

Hagadorn, J. W., R. H. Dott Jr., and D. Damrow. 2002. "Stranded on a Late Cambrian Shoreline: Medusae from Central Wisconsin." *Geology* 30 (2): 147–150.

Yochelson, E. L., and M. A. Fedonkin. 1993. *Paleobiology of* Climactichnites, *an Enigmatic Late Cambrian Fossil*. Smithsonian Contributions to Paleobiology 74. Washington, DC: Smithsonian Institution Press.

97. Newport State Park, Wisconsin

Dott, R. H., Jr., and J. W. Attig. 2004. *Roadside Geology of Wisconsin*. Missoula, MT: Mountain Press Publishing.

Kluessendorf, J., and D. G. Mikulic. 1989. "Bedrock Geology of the Door Peninsula of Wisconsin." In *Wisconsin's Door Peninsula: A Natural History*, edited by J. C. Palmquist, 12–31. Appleton, WI: Perin Press.

Paull, R. K., and R. A. Paull. 1977. *Geology of Wisconsin and Upper Michigan, Including Parts of Adjacent States*. Dubuque, IA: Kendall Hunt.

Schneider, A. 1989. "Geomorphology and Quaternary Geology of Wisconsin's Door Peninsula." In *Wisconsin's Door Peninsula: A Natural History*, edited by J. C. Palmquist, 32–48. Appleton, WI: Perin Press.

98. Two Creeks Buried Forest, Wisconsin

Black, R. F. 1970. *Glacial Geology of Two Creeks Forest Bed, Valderan Type Locality, and Northern Kettle Moraine State Forest*. Wisconsin Geological and Natural History Survey Information Circular 13.

Broecker, W. S., and W. R. Farrand. 1963. "Radiocarbon Age of the Two Creeks Forest Bed, Wisconsin." *Geological Society of America Bulletin* 74 (6): 795–802.

Prouty, C. E. 1960. *Paleozoic and Pleistocene Stratigraphy across Central Wisconsin*. Michigan Basin Geological Society Annual Field Excursion.

Wilson, L. R. 1932. "The Two Creeks Forest Bed, Manitowoc County, Wisconsin." *Transactions of the Wisconsin Academy of Sciences, Arts, and Letters* 27: 31–46.

99. Fossil Safari, Wyoming

Ambrose, P. D. 1996. *Fossil Butte National Monument: Along the Shores of Time*. Vernal, UT: Dinosaur Nature Association.

Grande, L. 1984. *Paleontology of the Green River Formation, with a Review of the Fish Fauna*. Geological Survey of Wyoming Bulletin 63.

Grande, L. 2013. *The Lost World of Fossil Lake: Snapshots from Deep Time*. Chicago: University of Chicago Press.

Lageson, D. R., and D. R. Spearing. 1988. *Roadside Geology of Wyoming*. Missoula, MT: Mountain Press Publishing.

100. Red Gulch Dinosaur Track Site, Wyoming

Lockley, M. 1991. *Tracking Dinosaurs: A New Look at an Ancient World*. New York: Cambridge University Press.

Lockley, M., and A. P. Hunt. 1995. *Dinosaur Tracks and Other Fossil Footprints of the Western United States*. New York: Columbia University Press.

Southwell, E. H., and B. H. Breithaupt. 1998. "Wyoming's Vertebrate Tracks: 130 Years of Discovery." *Journal of Vertebrate Paleontology* 18 (3): 79A.

Thulborn, T. 1990. *Dinosaur Tracks*. New York City: Springer Publishing Company.

101. Dinosaur Center, Wyoming

Chambers, P. 2002. *Bones of Contention: The* Archaeopteryx *Scandals*. London: John Murray Publishing.

Feduccia, A. 2012. *Riddle of the Feathered Dragons: Hidden Birds of China*. New Haven, CT: Yale University Press.

Shipman, P. 1998. *Taking Wing:* Archaeopteryx *and the Evolution of Bird Flight*. New York: Simon and Schuster.

Wellnhofer, P. 2009. Archaeopteryx: *The Icon of Evolution*. Munich, Germany: Friedrich Pfeil Publishing.

Index

Abo Formation, 132
Acadian orogeny, 140
Acadoparadoxides briareus, 45
Acrospirifer, 92
adaptive radiation, 186
Agassiz, Louis, 138, 166, 210
Agate Fossil Beds National Monument, 116–17
age dating, 104, 126, 182
Age of Crinoids, 10, 176
Age of Dinosaurs, 20
Age of Fishes, 3, 10, 74, 152, 162
Age of Mammals, 12, 20, 156
Age of Man, 13
Age of Reptiles, 12, 114, 146, 216
Age of Trilobites, 58
agnathids, 128
Alabama, 16–20
Alaska, 11, 22–23, 168, 200
Albertosaurus, 47, 216
alder, 22, 62
Aleutian Trench, 22
Alexander I Island, 5
algae, 8–9, 13, 54, 72, 84, 86, 96, 102–3, 120, 176, 178, 190, 208, 212
Alleghanian orogeny, 164
Allentown Dolomite, 130
alligators, 50, 174, 182
Allodesmus, 38
Allosaurus, 3, 11, 40, 42, 46, 188–89
Alpena limestone, 99–100
Alvaraz, Luis and Walter, 4
American Museum of Natural History, 3, 112
ammonites, 10, 45, 80, 106, 120–21, 172, 180
ammonoid, 28
amphibians, xii, 9, 10, 18, 19, 156, 162–63, 164
angiosperms, 62
Anning, Mary, 2
Antarctic Peninsula, 5

anthracite, 164
Anthropocene, 7, 14
anticline, 154–55, 197
Anza-Borrego Desert State Park, 32–33
Anza-Borrego Desert Paleontology Society, 32–33
Apex Chert, 8
Aphelaspis, 58–59
Aphelaspis brachyphasis, 59
Apatosaurus, 46, 188
Apex Chert, 8
Appalachian Mountains, 10, 68, 73, 76, 92, 100, 106, 126, 140, 164
Appalachian Plateau, 202
Araucarioxylon, 26
Arbuckle Anticline, 154–55
Archaeopteris, 198
Archaeopteryx, 4, 11, 188, 216–17
Archaeotherium, 157
Archean Eon, 7–8, 104, 130
Archelon, 146
Archimedes, 16–17, 28–29, 67, 202–3
Arcy-sur-Cure, 1
Ardmore (OK), 154
Argentinosaurus, 12
Arizona, 24–27, 34, 76
Arkansas, 28–31
armadillo, 13, 110, 166
arthropods, 130
Arundel Formation, 90
Asaphiscus wheeleri, 186–87
ash, 26, 34, 62, 81, 118–19, 124, 132, 158, 198, 200, 212
ashfall, 34, 118–19
Ashfall Fossil Beds State Historical Park, 118–19
Ash Hollow Formation, 118
Asterocyclina, 52–53
asteroid, 4, 120
Astrodon johnstonii, 90–91
Atrypa, 92–93
Aturia, 52–53

Augusta, 88
Aurora Fossil Museum, 142–43
Australia, 8, 124, 208

badger, 174
badlands, 26–27, 32–33
Bakersfield (CA), 38
bald cypress, 63
Bangor Limestone, 16, 17
barracuda, 50
basalt, 62
Basin and Range, 122, 124
Batostoma, 104–5
Battleship Rock, 132–33
bears, short-faced, 13, 110, 174
beaver, giant, 13, 50, 156, 166
beech, 62, 200
belemnite, 48–49, 128, 145
Belemnitella Americana, 48
Beneski Museum of Natural History, 94–95
Bering Land Bridge, 182
Berlin-Ichthyosaur State Park, 122
Besser Museum, 98–99
Big Bone Creek valley, 82
Big Bone Lick State Historic Site, 82–83
Big Brook Preserve, 128
Bigenerina, ix
birch, 63
birds, xii, 11–12; Cretaceous, 146, 148; earliest, 11, 188, 216–17; extinct, 60; flightless, 32, 60–61, 146; Miocene, 38, 118; Pleistocene, 36
bison, 37, 50, 82, 166, 180
bivalves, 24, 30, 44, 78, 106, 120–21, 151, 172–73, 192
Blackberry Hill, 206–7
Blasdell, 140
blastoids, 10, 16, 28, 202–3
Bluefield Formation, 203
body fossils, 4–6, 9, 18
Bone Valley, 50–51

241

bone wars, 40, 46
Borophagus, 33
Borrego Badlands, 33
Boston Mountains, 28
boulder star coral, 55
Bozeman, 114
Brachauchenius, 81
brachial valve, 24
brachiopods, xii, 5, 9, 10; Devonian, 74–75, 76–77, 92–93, 138–39, 140, 152–53, 160–61, 194–95, 205; Mississippian, 16, 28–29, 202; Paleocene, 52; Pennsylvanian, 24–25, 78–79, 132–33, 184–85; Ordovician, 72–73, 83, 104, 150–51, 154–55
Brachiosaurus, 188
Braidwood biota, 68
brain coral, 55
Brevard Museum of History and Natural Science, 56–57
Bromide Formation, 154–55
Brontosaurus, 42, 46
Brown, Barnum, 112
Browning Park, 106–7
Bruneau-Jarbidge, 118
bryozoans, xii, 5, 9, 10, 84–85, 104–5; Archimedes, 16–17, 28–29, 67, 202–3; Ordovician, 72, 88–89, 104–5, 150, 154; reefs of, 60–61, 84–85, 190
Burgess Shale, 68, 188
Burlington Limestone, 110
burrows, 6, 18, 19, 26, 116, 136

Caesar Creek State Park, 150–51
Calamites, 164
calcite, 5
calcium carbonate, 5, 16, 30, 52, 54, 70, 76, 78, 96, 130, 208, 212; dissolution of, 174; sand of, 60; in soil, 178
California, 24, 32–39, 158
Calymene celebra, 66–67
Calymene niagarensis, 67
calyx, 45, 79, 176, 184, 202–3
Camarasaurus supremus, 40–41
Cambrian explosion, 8–9, 84, 162, 190, 206
Cambrian Period, 3, 7–9, 58, 69, 120, 130, 154, 186, 196, 206
camels, 12, 13, 32, 36, 37, 63, 64, 108, 116, 118, 174, 180, 182
Cane River Formation, 86
Cannonball Formation, 170

Cañon City, 40, 46
carbonate. *See* calcium carbonate
Carboniferous Period, 10–11
carbonization, 5
Carcharodon megalodon, 142, 143, 144
Caririchnium, 136
carnivores, 11, 32, 36, 94, 110, 142, 172, 188
Carriza Badlands, 32
Cascade Range, 158, 198
casts, 4, 92–93, 95, 124, 158–59, 163, 202
catastrophism, 2–3
cats. *See* saber-toothed cat
caves, 33, 52, 60–61, 110–11, 174
cedar, 62
centipedes, 152
Central Valley, 38
cephalon, 58–59, 66, 186–87, 204–5
cephalopod, 5, 74, 85, 121, 140–41, 160, 190–91
Cerro Gordo Member, 76
chain coral, 208–9
chalk, 6, 20–21, 80, 178
Champlain, Lake, 190
Champlain Black, 88
Chazy Fossil Reef National Natural Landmark, 190–91
chert, 8, 102–3, 108, 197
Chesapeake and Delaware Canal, 48–49
Chesapecten jeffersonius, 192–93
Chickaloon Formation, 22
Chicxulub crater, 12
Chinle Formation, 26–27, 134–35
Chippokes Plantation State Park, 192–93
chitinous, 5
Chonetes, 78–79
Chonetina, 184–85
Cincinnatian Series, 150
Cincinnati School of Paleontology, 9, 150
Cincosaurus cobbi, 19
clams, 42, 44, 68, 78, 88, 120, 178, 188, 192, 208
Clark, William, 82
Clarkia Fossil Bowl, 62–63
clay, 4, 30, 62, 64, 90, 168, 173, 174, 178, 180, 186, 210
Clayton Formation, 20
Clayton Lake State Park, 136–37
Clidastes tortor, 80
Climactichnites, 207
climate, 5–6, 10, 12, 13, 14

Climatograptus, 196
Clovis culture, 108
club moss, 68, 164
Clypeaster, 53
coal, 10, 13, 18–19, 23, 28, 34, 68, 164, 198
coelenterates, 88
Coelophysis, 134–35
Coeymans Formation, 138–39
Coffee Sand, 106
colonial organisms. *See* bryozoans; corals; graptolites
Colorado, 13, 40–45, 46, 134, 136
Colorado Plateau, 134
Columbian mammoth, 82, 182–83
columnals, 79, 88, 99, 133, 140, 176–77, 184–85, 202
Comanche Peak Formation, 178
Composita, 24–25, 78–79, 132–33
Composita elongata, 78
Composita ovata, 78
Composita subtilita, 24–25, 78
Composita trilobata, 78
Conasauga Formation, 58
concretions, 68, 106–7, 171
Condon, Thomas, 156
conglomerate, 26, 96–97, 164
conifers, 10, 11, 26, 90, 210
Connecticut, 40, 46, 90, 95
Coon Creek Science Center, 172–73
Cope, Edward, 3, 40, 46, 80, 114
coprolites, 2, 6, 110, 128–29, 142–43, 180
corals, xii, 5, 9–10, 190; brain, 55; branching, 205; chain, 208–9; colonial, 9, 54, 74–75, 76–77, 100–101, 190, 208–9; contemporary, 54; extinctions of, 152; honeycomb, 104, 208–9; horn, 16–17, 70–71, 73, 78–79, 98–99, 104, 161, 208–9; lace, 17; Pleistocene, 54, 60; rugose, 70–71; solitary, 88; star, 55
Cordell Formation, 209
Cordova Shell Limestone, 88
corkscrews, 16, 17, 28, 202, 203
Corys Lane, 164–65
Costispirifer, 92
cow sharks, 39
Coyote Lake, 22–23
crabs, 11, 122, 172, 173
Cretaceous Period, 12, 48–49; climate of, 5; dinosaurs of, 11, 90–91, 112, 114–15, 136–37; index fossils of, 49, 106, 178; K-Pg extinction, 20–21, 48;

Mississippian Embayment of, 30, 120, 172–73; marine life of, 180; Western Interior Seaway of, 81, 136–37, 146, 148–49
crinoids, xii, 10, 11, 29, 79, 81, 83, 88–89, 98–99, 104–5, 132–33, 176–77, 184–85
crocodiles, 11, 26, 38, 90, 95, 122, 134–35, 144–45, 156, 172, 180, 212
Cro-Magnon, 1
Crown Point Formation, 88
crustaceans, 39, 207
Crystal Forest, 26
Ctenacanthus, 44
Cuvier, Georges, 2
cyanobacteria, 8, 96–97, 130, 138, 190
cycads, 10, 18, 26
cypress, 63, 170, 198, 199, 200, 212
Cyrtospirifer whitneyi, 77

Daemonelix, 116–17
Daemonelix Trail, 116–17
Dakota Group, 40
Danville (PA), 160
Danville Bryozoan Reef, 84
Darwin, Charles, 1, 102, 216
Daspletosaurus, 115
da Vinci, Leonardo, 1
dawn redwoods, 62–63, 200
Deccan lava flows, 12, 20, 48
Decorah Shale, 104–5
deer, 13, 118
Delaware, 48–49, 82
Demopolis (AL), 20
Demopolis Formation, 106
Denver Museum of Nature and Science, 44–45
Desmostylus, 39
Devils Backbone, 28–29
devil's corkscrew, 116
devil's toenails, 106, 120, 178
Devonian Fossil Gorge, 74–75
Devonian Period, 10, 70, 74, 76, 98, 92, 100, 138, 140, 152, 160, 162, 194, 196, 204
diatoms, 36
Dilophosaurus, 188
dinomania, 3–4, 46, 114
Dinosaur Center, 216–17
Dinosaur Lake, 42
Dinosaur National Monument, 188
Dinosaur Park, 90

dinosaurs, xii, 11–12; discoveries of, 40, 80, 90, 112–13, 188–89; early, 44, 134; eggs of, 115, 217; extinction of, 4, 20, 114; museums featuring, 46–47, 80–81, 114–15, 146–47, 216–17; quarries of, 40–41, 188; skin of, 146–47; teeth of, 91; trackways of, 42–43, 94–95, 136–37, 214–15. *See also* bone wars; dinomania; Dinosaur Park; Dinosaur National Monument; Wyoming Dinosaur Center
Diplodocus, 188–89
diplodocus, 188–89
Diplomystus, 212–13
Diploria, 54–55
dire wolves, 13, 36
Discorbis, ix
Dissected Till Plains, 118
dissolution, 4, 168, 174
dogs, 12, 13, 32, 33, 64, 116, 156
dolomite, 66, 130–31, 154, 174
dolphins, 38, 50, 142, 143, 144, 166
Double Roadcut, 16
Douglas fir, 198
duck-billed dinosaurs, 106, 114
Ducrotay de Blainville, Henri Marie, 1
dugong, 13, 50–51, 142
Dunkleosteus, 44, 74–75, 152, 160, 172
Dunkleosteus terrelli, 44

Earth: age of, 1–3; atmosphere of, 8, 9
earthquakes, 4, 22, 122, 160
East African Rift System, 96
echinoderms, 5, 16, 72, 84, 110, 176
echinoids, 52, 178, 179, 180
Ecphora quadricostata, 193
Ediacara Hills, 8
Edisto Beach, 166–67
Edisto fauna, 166
Edmontosaurus, 146
Edwards Formation, 178
Effigia, 134
Elasmosaurus, 46
Elephant Knees Mesa, 32–33
elephants, 80, 82, 108, 118, 174. *See also* mammoths; mastodons
Elrathia kingii, 186–187
elm, 22, 170, 198
Enchodus, 181
Encrinurus, 67
Entrada Sandstone, 134–35
Eocene, 22, 52, 86, 156, 200–201, 212

Eosphaera, 103
epicontinental, 98, 148
Equus simplicidens, 65
Eridophyllum, 70–71
erosion, 11, 74, 130, 174, 177, 180
Essex biota, 68
Essexella asherae, 69
Eucalycoceras, 121
Euproops danae, 69
Eusthenopteron, 44
evolution, 3, 5, 6
Exogyra, 30, 49, 106–7, 120, 180
Exogyra cancellata, 30
Exogyra costata, 49
Exogyra ponderosa, 30, 106–7
exoskeleton, 5, 9, 58, 66, 70, 78, 98, 100, 140, 160, 186, 204–6
extinction events, 9, 10, 11, 12, 13, 20, 26, 34, 48, 60, 114, 120, 152, 160, 184

Falls of the Ohio, 70–71
faults, 22, 200
Favosites, 138, 208–9
Fayetteville Formation, 28
Fenestella, 17
fern, 5, 10, 12, 22–23, 26, 42, 68–69, 90, 198, 212
Field Museum of Chicago, 4
fish, xii, 6, 12, 80, 128, 212–13; jawed, xii, 10, 44; jawless, xii, 9, 10, 50, 162. *See also* Age of Fishes; *Dunkleosteus*; lobe-fin; placoderms; ray-fin; sharks; *Xiphactinus*
Fisk Quarry Preserve, 190
flint, 108
Florida, 30, 50–57
Florida Caverns State Park, 52
Florida East Coast Railway, 54
Florida Keys, 54
Florissantia, 200–201
flowstone, 52
Fluvanna roadcut, 178–79
folds, 24, 195
foraminifera, 5, 10, 20, 52–53, 86–87
forests, fossil, 22, 26–27, 126–27, 158–59
Fort Peck Interpretive Center, 112
Fossil and Prairie Park Preserve and Center, 76
Fossil Butte National Monument, 212
fossil fuels, 10, 13, 18, 36, 38, 164, 196
Fossil Lake, 212
Fossil Park, 152–53

fossils: body, 4–6; definition, 4; index, 5; trace, 6
Fossil Safari, 212–13
Franklin, Benjamin, 82
Frankstown Sand, 106

gastropods, 5, 16, 24, 76, 88, 132, 152, 160, 180–81, 190
geologic time, xii, 3, 7
Georgia, 58–59
Ghost Ranch, 134–35
Ghost Ranch Museum of Paleontology, 134–35
Giant Logs, 26–27
Giganotosaurus, 216
Gilead (NE), 120
Ginkgo adiantoides, 199
ginkgoes, 26, 68, 199
Ginkgo Petrified Forest State Park, 198–99
glacial polish, 130
glaciation, 8–11, 138, 174, 210
Glen Lyn roadcut, 202–3
Glyptagnostus, 58
Gondwanaland, 28, 162, 164
Goodsell Ridge Preserve, 190–91
grabens, 94, 200
Grafton Quarries, 66
Grand Isle, 86
Graneros Shale, 120
graptolites, 3, 9–10, 72, 152, 154, 196
Gray Fossil Site, 174–75
Great American Biotic Interchange, 13
Great Barrier Reef, 208
Great Flood of 1993, 74
Great Hall of Dinosaurs, 46
Great Ordovician Biodiversification Event, 154
Greenhorn Limestone, 121
Green Mill Run, 144–45
Greenops, 140–41
Green River Formation, 6, 44, 212
Grewingkia canadensis, 73
ground sloths, 13, 32, 36, 37, 50, 64, 82, 108, 166, 172, 180
Gryphaea nebrascensis, 215
Gryphaea vesicularis, 30
guards, 48
Gunflint Chert, 102–3
Gunflintia, 102–3
Gunflintia minuta, 103
gymnosperms, 62

Gypidula, 138–39
Gyrodes major, 181

Hadean Eon, 7
Hagerman Fossil Beds National Monument, 64–65
Hagerman horse, 64–65
Hall, James, 138
Hallopora, 104
Halysites, 208–9
Hamburg stromatolites, 130–31
Hancock Park, 36
Hangenberg event, 152
Hanging Rock Iron Region, 90
Harlan's ground sloth, 82, 108
Hawaii, 60–61
Hebertella, 151
Helderberg Escarpment, 138–39
Heliophyllum, 70–71
Hell Creek Formation, 112–13, 146
hemlock, 124, 126, 170, 210
herbivores, 36–37, 42, 47, 94, 166, 174, 188–89
Hesperornis, 146
Hexagonaria, 74–75, 100–101
Hexagonaria percarinata, 100–101
hippopotamus, 39, 82, 118
Hitchcock, Edward, 94, 214
Hitchcock Ichnological Cabinet, 94
Hogue Creek, 194–95
holaspid, 66
Holocene, 13, 56, 60, 158, 166
Holocene Maximum, 13
holotype, 216
hominids, 13
Homotelus bromidensis, 155
honeycomb corals, 209
Hooke, Robert, 2
horn corals, 16–17, 70, 73, 78–79, 98–99, 104, 148, 161, 208–9
horse chestnut, 198
horses, 12, 13, 37, 50, 64–65, 82, 108, 110, 118, 166, 180
horseshoe crabs, 18, 45, 68, 69, 206
Horseshoe Harbor, 96–97
horsetails, 10, 18, 22, 26, 68, 164
Hot Springs (SD), 168
Howellella, 138
Hubble Space Telescope, 92
Hudson Bay, 7
humans, 6, 13, 56

Hustedia, 79
hydrocarbons, 13
Hyneria, 162–63
Hynerpeton, 162–63
Hystriculina, 132

Iapetus Ocean, 76, 88–89, 190
ice age, 13–14, 36, 72, 108, 110–11, 130, 166, 182–83, 210
ichnofossil, 6, 110, 136
ichnology, 94, 214
Ichthyosaurus, 2, 11, 81, 122–23
Idaho, 62–65, 118–19, 198
igneous, 7, 104, 126, 132, 158
Iguanodon, 136–37
Illinois, 30, 66–69, 126, 172–73
impressions, 4–6, 59, 62, 69, 80, 92, 146, 158–59, 164, 187, 188. *See also* trackways
index fossils, 5, 16, 49, 58–59, 72–73, 79, 93, 106, 178, 194, 196
Indiana, 9, 70–73, 150, 208
Indiana Limestone, 88
Indian Gardens, 24
Inoceramus, 120–21
insects, 6, 10, 18, 34, 36, 68, 156, 162, 164, 200–201, 212
invertebrates, 5
Iowa, 74–77, 104
iridium, 4, 12
Iron Age, 5, 56
iron ore, 90
iron oxide, 59, 124
Isle La Motte, 190
Isthmus of Panama, 32
Isua sequence, 7
Isurus hastalis, 39
Itagnostus interstrictus, 186–87

James River, 192
Jasper Forest, 26
jawbones, 173, 181
Jefferson, Thomas, 33, 82, 192
Jefferson's Chesapeake scallop, 192
Jefferson's ground sloth, 82
Jeffersonville Limestone, 70
jellyfish, 68, 69, 88, 206–7
Jemez Mountains, 132
Jemez Springs Shale, 132
John Boyd Thacher State Park, 138
John Day Fossil Beds National Monument, 156–57

Jurassic Period, 11, 40, 42, 46, 106, 114, 188, 214, 216

Kakabekia, 102–3
Kakabekia umbellata, 102–3
Kansas, 44, 78–81
Kanwaka Formation, 78–79
karst, 52, 110, 174
Kaskaskia Sea, 76, 100, 140, 160–61
Kasota Limestone, 88
Kauai, 60–61
Kauai mole duck, 60–61
Kellwasser event, 152
Kentucky, 9, 72, 82–85, 150
Keweenaw Peninsula, 96
Key Largo Limestone, 54
Keyser Formation, 92–93
Kimmswick bone beds, 108
Knightia eocaena, 212
K-Pg extinction, 20

La Brea Tar Pits, 36
La Brea Woman, 36
Ladonia Fossil Park, 180–81
Lafarge Fossil Park, 98–99
lagerstätte, ix, 62, 68, 126, 172, 186, 212
Lambeophyllum, 88
lampshells, 154, 194
land bridge, 13, 32, 157, 166, 182, 200
Lapworth, Charles, 3
Laramide orogeny, 12
Laramidia, 120
Laurasia, 16, 28, 130, 162
Laurentia, 164, 190
lava, 11–12, 48, 60, 96, 158–59, 198
Lava Cast Forest, 158–59
leaves, 9, 23, 60, 62–63, 164–65, 198–99, 200–201, 212
Leperditia, 138
Lepidocyclina, 52
Les Eyzies, 1
Liberty Formation, 72, 73
Lime Creek Formation, 77
limestone, 10–11, 54, 88–89. *See also* calcium carbonate; reefs
limonite, 90
Lincolnshire Formation, 197
lion, American, 36
Lithophaga, 54–55
Little Blue River outcrop, 120
Little Cedar Formation, 74

Little Ice Age, 13
lizards, 11, 122. *See also* dinosaurs
llamas, 13, 50, 180
Llano Estacado, 178
lobe-fins, 44, 162
Long Logs, 26
Lophophyllidium, 78–79
Lost River Quarry, 204–5
Louisiana, 34, 86–87
Lower Peninsula, 98, 100, 208
Ludwig Leichhardt, 2
Lyell, Charles, 3, 138

Maastrichtian age, 146
Maclurites, 88–89
Maclurites magnus, 89
Madera Group, 132–33
magnetite, 90
magnolia, 22, 198, 200
Mahantango Formation, 160, 195
Maha'ulepu Valley, 60
Maiasaura, 217
Maine, 18, 88–89
Makauwahi Cave, 60–61
mammals, xii, 11, 12–13, 36, 50, 64, 82, 108, 110, 116, 118, 156, 168, 174, 182
mammoths, 2, 5, 13, 32–33, 82, 168, 169, 182–83; teeth of, 37, 50–51, 109, 180
Mammoth Site, 168–69
Mammut americanum, 108
Mammuthus columbi, 168–69, 182–83
Mammuthus meridionalis, 32
Manlius Limestone, 138–39
Mantell, Gideon, 214
maples, 63, 126
marble, 126
Marginulina, ix
marl, 20, 30–31, 172
Marlbrook Marl, 30–31
Marsh, Othniel C., 3, 40, 46, 114
Marsh-Felch Quarry, 40–41
Maryland, 90–93, 160, 192
Mary Lee Seam, 18
Massachusetts, 94–95, 138, 164
mastodons, 4, 12–13, 32, 36–37, 50, 82–83, 108–9, 166–67, 180
Mastodon State Historic Site, 108–9
Mazon Creek, 68
Mazonia-Braidwood State Fish and Wildlife Area, 68
Mediospirifer, 161

Megalodon, 13, 39, 51, 142, 143, 144, 167
Megalonyx, 33
Megalosaurus, 114
Meganeura, 44
Megastrophia, 153
Menoceras, 44, 116–17
meraspid, 66
Meristella, 92–93
metamorphism, 24, 96, 126
meteorites, 12, 20–21, 48, 114, 120, 152, 160
Michigan, 56, 96–101
Michigan Basin, 208
Micraster, 6
microfossils, 8, 103, 110
Microraptor, 12, 217
microtektites, 20–21
Midcontinent Rift System, 96
millipedes, 18–19, 110–11, 152
Mine Lot Falls, 138–39
mineralization, 5
Mineral Wells Formation, 184
Mineral Wells Fossil Park, 184–85
Minkin Footprints, 18
Minnelusa Formation, 168
Minnesota, 74, 90, 102–5
Mint Spring, 196–97
Miocene, 32, 38, 50, 62, 116, 118, 124, 142, 144, 156, 174, 192, 198
Mississippi, 106–7
Mississippi Embayment, 30–31, 172
Mississippian, 10, 16–17, 28, 88, 110, 176, 196, 202–3
Missouri, 108–11
Missourium kochii, 108
moa-nalo, 60
molars, 32, 36, 50, 82, 83, 109, 180
molds, 4, 88, 92–93, 120, 124, 158–59, 179, 181
mole duck, 60–61
mollusks, 10, 180, 192–93. *See also* ammonites; bivalves; cephalopods; clams; gastropods; oysters
molts (molting), 5, 58, 66, 98, 186, 187, 204–5
Montana, 80, 112–15, 217
Montastraea, 54–55
Montastraea annularis, 54–55
Monterey Formation, 36, 38
Montour Preserve, 160–61
Moodys Branch Formation, 86

245

Morozovella, 20
Morrison Formation, 42, 188
mosasaurs, 11–12, 30–31, 80–81, 120, 144–45, 148–49, 172–73, 180–81
Moscow Landing, 20–21
Moses Maimonides, 1
moss animals, 84–85, 88
Mount Simon Sandstone, 206
Mucrospirifer, 153, 194–95
mud cracks, 130–31, 136–37
mudflats, 78, 136
mudflows, 59
mudstone, 36, 59, 96, 112, 170, 180
Murchison, Roderick, 3
Museum of the Rockies, 113, 114–15
musk ox, 82, 180

Naco Formation, 24
Naco Paleo Site, 24
Narragansett Bay, 164
National Museum of Natural History, 88
National Museum of the American Indian, 88
National Museum of Women in the Arts, 88
Natural History Museum of Los Angeles, 112
nautilus, 52, 53
Nebraska, 116–121, 182
Nemograptus, 196
nematodes, 100, 152
Neogene Period, 13, 156
Neospirifer, 132
Neuropteris, 69
Nevada, 13, 38, 84, 122, 124, 158
Newberry National Volcanic Monument, 158
New Hampshire, 126–27
New Jersey, 31, 128–31, 160
New Mexico, 11, 80, 132–37, 178
Newport State Park, 208–9
New York, 13, 76, 106, 110, 112, 138–41, 160
North Carolina, 64, 142–45
North Dakota, 34, 146–49, 170
North Dakota Heritage Center and State Museum, 146

oak, 22, 62
obsidian, 108
Odiorne Point State Park, 126–27
Odontopteris, 164–65
Ohio, 9, 70–72, 90, 150–53, 208
Ohio River, 70
oil seeps, 36–37
Okanogan Highlands, 200
Oklahoma, 93, 154–55
Oligocene, 52, 156
oolites, 130
opal, 124–25
open-pit mine, 18
Orcinus orca, 172
Ordovician Period, 9, 72, 82, 84, 88, 104, 130, 150, 154, 172, 178, 184, 190, 196
Oregon, 64, 80, 156–59
Oriskany Sandstone, 92–93
ornithischians, 188
ornithopod, 136
ostracod, 138
Ostrea falcata, 30–31
Ouachita Mountains Province, 28
Overseas Railroad, 54
Oxoplecia gouldi, 155
oysters, 30–31, 32, 49, 78, 88, 106–7, 120–21, 178–79, 180, 215
Ozark Plateaus Province, 28

Pachycephalosaurus, 112
Pachyphyllum woodmani, 76–77
Paibian, 58
Palaeocastor, 116–17
Paleocene, 20, 22, 52, 170, 200
Paleocene-Eocene Thermal Maximum, 200
paleoclimate, 156
Paleogene Period, 12–13
paleogeography, 5, 106
Paleo-Indians, 168
paleontology, 1–4
panda, red, 174–75
Pangaea, 10–11, 26, 28, 30, 42, 94, 96, 106, 122, 132, 134, 152, 154, 164, 202
Panthalassa Sea, 10, 122
Payson, 24
Peabody Museum, 46–47
Peace River, 50
peat, 5, 22, 56, 60, 126, 164, 210
peccary, 110, 111, 166, 174
Pecopteris, 164–65
Pecten jeffersonius, 192
pedicles, 24–25, 153
pelecypod, 48, 72, 88, 140, 160, 180, 184
Pembina Gorge, 148–49

Penn Dixie Fossil Park and Nature Reserve, 140
Pennsylvania, 36, 90, 92, 160–63
Pennsylvanian, 10, 18, 24, 28, 68, 78, 132, 154, 164, 184
Pentremites, 16–17, 203
Pere Marquette State Park, 66
Permian Period, 10–11, 26, 44, 133, 150, 165, 184, 194, 202
permineralization, 5
Peronopsis interstricta, 186–87
Petoskey State Park, 100
Petoskey stones, 100–101
Petrified Forest (CA), 34–35
Petrified Forest National Park (AZ), 26–27
petrified forests, 26, 34, 198
petrified wood, 2, 5, 23, 26–27, 124–25, 157, 198–99
Petrified Wood Park, 170–71
Phacops, 140–41, 153, 204–5
Phacops rana, 153
Phillips, John, 3
photosynthesis, 8, 97
phylum, 9, 84, 88, 100, 176
Picket Wire Track Site, 42–43
Pierre Shale, 148–49
Pila's palila, 60
Pilbara Group, 8
pinecones, 22
Pitkin Limestone, 28
placoderm, 10, 162, 172
plankton, 9, 48, 98, 196
plants, 5, 8, 9, 10, 22, 200–201. *See also* algae; leaves; petrified wood; *specific plant names*
Platecarpus, 81
plate tectonics, 4
Platyceramus platinus, 44
Platypterygius, 81
Platystrophia, 72–73, 151
Platystrophia acutilirata, 72
Pleistocene, xii, 13–14, 32, 36, 50, 54, 82, 108, 110, 126, 166, 180, 182, 210
Plesiosaurus, 2–3, 11–12, 120, 180
Pliocene, 32, 34–35, 50, 64, 142, 144, 174, 192
pliosaurs, 81
Pottsville Formation, 18
Prairie Bluff Chalk, 20
Prasopora conoidea, 105
Precambrian Era, 7

Principle of faunal succession, 2, 5
Principle of fossil correlation, 5
Principle of superposition, 2
Priscacara liops, 213
Pristinailurus, 175
proboscidean, 32, 168
Prognathodon overtoni, 172
Proterozoic Eon, 8
pseudomorph, 5
psittacosaurs, 47
Pterotrigonia thoracica, 173
Punctospirifer, 132
Pungo River Formation, 142
Purgatoire River, 42
Pycnodonte mutabilis, 49
pygidium, 58, 66, 186–87, 204
pyrite, 5, 140
pyritication, 140

quartz, 26, 35, 92, 197
quartzite, 126
Quaternary Period, 13–14, 56, 60, 158. *See also* Pleistocene

rabbits, 12
Racemiguembelina, 20
Radio Black, 88
radiocarbon dating, 3, 126, 210
Rafinesquina, 151
Rainbow Forest Museum, 26–27
raindrops, 19, 95, 206
Rancholabrean fauna, 166
ray-fins, 10, 152, 162
Red Gulch Dinosaur Track Site, 214–15
Red Hill, 162–63
redwoods, 34, 62–63, 198, 200
reefs, 9–10, 54, 60–61, 84, 100, 160, 190, 208
reptiles, xii, 5, 11–12, 18–19, 48, 81, 128, 134, 136, 148; marine, 28, 46, 122–23, 172. *See also* dinosaurs
Republic graben, 200
resin, 4
Rheic Ocean, 16
Rhipidomella, 92–93, 205
rhinoceros, 12, 13, 44, 116–17, 118–19, 156, 174–75
Rhode Island, 22, 164–65
Rhode Island Formation, 165
Richmond (IN), 72
Richmondian, 72

rift, 13, 94, 96, 154
Rim Fire, 158
ripple marks, 42, 136, 207, 214
Riverbluff Cave, 110–11
roche moutonnée, 130, 131
rock dogs, 66
Rockford (IA), 76
Rock Island State Park, 176–77
Rocky Mountains, 12, 116, 132
Rodinia, 8, 96, 190
Rooster Bridge, 20
Round Mound, 78–79
rugose corals, 10, 70–71

saber-toothed cats, 13, 32, 36–37, 64, 110, 118, 157, 172, 174, 182
Salem Limestone, 88
Salesville Shale, 184
Salton Trough, 32
sandstone, 92–94, 170–71, 188–89, 194, 200, 202; impressions in, 6, 8, 136, 194, 206
Sandy Mile Road, 92
Saratoga (AK), 30
sassafras, 198, 201
sauropods, 90, 188
Sauroposeidon, 12
Scapanorhynchus, 107
Schilderia, 26
scorpions, 68, 152
sea buds, 16, 202
sea level changes, 30, 54, 68, 126, 208
sea lilies. *See* crinoids
Sedgwick, Adam, 3
sedimentary rock, 7, 104
sedimentation, 104
seed ferns, 18, 69, 164, 165
seedpods, 22, 63
sequoia, 22, 34, 170
Sequoia langsdorfii, 34
shale, 78–79, 140–41, 150–51, 160, 194–95, 196–97, 200–201; oil in, 13, 36; tracks in, 18–19, 94–95, 136; weathering out of, 24, 26, 28, 76, 153, 202. *See also* Burgess Shale; Decorah Shale; Graneros Shale; Jemez Springs Shale; Pierre Shale; Salesville Shale; Wheeler Shale
sharks, xii, 2, 10, 13; coprolites, 128–29, 181; Cretaceous, 48, 106–7, 128–29, 172, 180–81; Miocene, 38–39, 50–51, 142–43; Paleozoic, 10, 16, 28, 44, 162, 184; Pliocene, 144–45; Pleistocene, 166–67
Sharktooth Hill, 38–39
Shonisaurus, 122–23
Siberia, 4, 11, 58, 168
siderite, 68, 90
silica, 5, 9, 26–27, 34, 124, 152–53, 198
Silica Formation, 153
siltstone, 58, 132, 134, 160, 189
Silurian Period, 3, 9–10, 208
sinkhole, 52, 60–61, 168, 174
Siphonophrentis, 71
slate, 126, 197
Smilodon populator, 172
Smith, William, 2, 5, 194
snails, 1, 11, 68, 152, 178, 181, 193, 209, 210
Snowball Earth, 8
soda straws, 52, 111
Solnhofen Limestone, 188
South Australia, 7–8, 124
South Carolina, 166–67
South Dakota, 114, 168–71
speleothem, 111
spicules, 5
Spinosaurus, 172
Spirifer, 75
Squalicorax, 107, 129
stag-moose, 82
stalactites, 52, 111
stalagmites, 111
star corals, 55
Stegosaurus, 11, 44, 46
steinkern, 178–79
Steno, Nicolas, 2
Sternberg, Charles, 80
Sternberg Museum of Natural History, 80–81
Steven C. Minkin Paleozoic Footprint Site, 18
Stone Age, 54
Stonerose Interpretive Center and Eocene Fossil Site, 200
Stromatocerium, 88
stromatolite, 8, 96–97, 102–3, 130–31, 138, 190
stromatoporoid, 138, 190–91
Stropheodonta, 92
Strophomena, 72, 155
Strophomena nutans, 72
Strophonella, 92
Stylemys, 157

Sue, 4
Sundance Formation, 214
Sundance Sea, 11, 214
supercontinents, 8, 10, 42, 94, 106, 122, 134, 154, 164, 190, 202
Supersaurus, 216
sweetgum, 198
sycamore, 62, 170
Sylvania, 152–53

Taconic orogeny, 72
Talpanas lippa, 61
tapir, 50, 174
Tawa, 134
tectonics, 4, 96
teeth, 4, 5, 6, 13. *See also* molars; sharks
Teleoceras, 175
Temblor Formation, 38
Tennessee, 172–76, 190
Tenontosaurus, 115
Tentaculites, 138–39
Tetraphalerella neglecta, 72–73
tetrapods, 18–19, 50, 162–63, 202
Texas, 34, 178–85
Texas Panhandle, 178
thecas, 83
theropod, 134, 136–37, 188, 214, 216
Thistlethwaite Falls, 72
Thomas Condon Paleontology Center, 156
thorax, 58, 66, 123, 186–87, 204–5
Tibbs Bridge, 58–59
till, 118, 210–11
timescale, 3, 7
Tippecanoe Sea, 72, 84, 92
Titusville, 56
Tombigbee River, 20
tortoises, 32, 118, 157, 182
trace fossils, 4, 6, 55, 88, 124–25, 170. *See also* burrows; coprolites; ichnofossils; impressions; steinkerns; trackways; tree molds
Trachypora, 205
trackways, 6, 18–19, 42–43, 94–95, 136, 202, 206–7, 214–15
tree molds, 158–59
T. rex. See Tyrannosaurus rex
Triassic Period, 11, 26, 28, 94, 122, 134
Triceratops, 3, 112–13, 114, 146–47, 180, 216
trihorned, 11, 112
trilobites, xii, 1, 3, 5, 9, 45, 58–59, 66–67, 73, 98, 130, 140–41, 150–51, 152–53, 154–55, 161, 186–87, 204–5; extinction of, 10
tsunamis, 20–21, 60, 122, 210
Tullimonstrum gregarium, 68
turbidity current, 141
Turritella trilira, 181
turtles, 11, 38, 50, 51, 90, 106, 166, 110, 146, 148, 172, 193
Twin Falls, 177
Two Creeks Buried Forest, 210–11
Tylosaurus, 181
Tyrannosaurus rex (T. rex), 4, 11, 112–13, 114–15, 146, 172, 188

U-Dig Fossils quarry, 186
Uintacrinus, 81
unconformity, 132, 206
uniformitarianism, 3, 138
Union Chapel Mine, 18
University of Wisconsin Geology Museum, 206
Utah, 13, 134, 186–89

Valley and Ridge Province, 202
valves, 24–25, 28, 78, 153, 160, 178, 184, 202
Vancleavea, 134–35
Vermont, 88, 138, 190–91
vertebrae, 30, 31, 39, 48, 123, 149, 188
vertebrates, xii, 9–12
Virgin Valley, 124–25
Virginia, 30, 33, 90, 92, 192–97

Waco Mammoth National Monument, 182–83
Wall of Bones, 188–89
Walnut Clay, 178
Wanakah Formation, 140
Wangs, 104–5
Washington, 34, 108, 118, 198–201
Washington, DC, 88
Waynesville Formation, 72–73
weathering, 86, 93, 108, 113, 195, 197–98
Western Interior Seaway, 120, 136, 146, 147, 148, 149
West Virginia, 195, 202–5
whales, xii, 12, 38, 39, 50, 142, 144, 145, 166, 172
Wheeler Shale, 186–87
Whitewater Formation, 72–73
Whitewater River Gorge, 72
Windley Key Fossil Reef Geological State Park, 54–55
Windom Shale, 140–41
Windover village, 56
Wisconsin, 67, 206–11
Wisconsinan glaciation, 210
W. M. Browning Cretaceous Fossil Park, 106
Woodworthia, 26
woolly mammoth, 5, 168–69
worm borings, 34–35
Wyoming, 6, 13, 45–46, 80, 126, 132, 212–17
Wyoming Dinosaur Center, 216–17

Xiphactinus, 80, 146–47

Yucatán Peninsula, 12, 20

Zaphrentis, 209
zebras, 64–65
zircon, 188
zooids, 84–85, 196
zooplankton, 55

Born in Ohio, **Albert B. Dickas** earned BA and MA degrees from Miami University (Oxford, OH). After serving in the US Navy, he obtained a PhD at Michigan State University and then joined Magnolia Petroleum (Mobil Oil) in the offshore salt dome province of Louisiana. From there he moved to Standard Oil of California, where he specialized in hydrocarbon exploration in the subsurface submarine canyons of the Sacramento Valley. Having satisfied his strong belief that firsthand industrial experience is a prerequisite for work in academia, he joined the faculty of the University of Wisconsin–Superior, where he taught for thirty-one years. He also founded an environmental research center and became intensely involved in the industrial exploration for oil in the Precambrian strata of the Midcontinent Rift System. During this time he led numerous field conferences, authored or coauthored more than forty-three papers, and delivered presentations from Nova Scotia to Siberia—all on the subject of Precambrian oil and Precambrian rifting. He is the author of *101 American Geo-Sites You've Gotta See* and *Ohio Rocks! A Guide to Geologic Sites in the Buckeye State*. Today he lives on the crest of Brush Mountain in southwest Virginia, where he continues to engage in research and plan travel excursions in his quest for new and interesting geo-sites on all seven continents.